AF553779

DICTIONARY OF BOTANY

DICTIONARY
OF
BOTANY

R.K. TANDON

A P H PUBLISHING CORPORATION
4435-36/7, ANSARI ROAD, DARYA GANJ
NEW DELHI-110 002

Published by
S.B. Nangia
A P H Publishing Corporation
4435-36/7, Ansari Road, Darya Ganj
New Delhi-110002
Ph. 23274050

Email : aphbooks@gmail.com

2021

Printed at :
Balaji Offset
Navin Shahdara,
Delhi-32

Preface

Botany is a branch of biology which involves the study of plants. It is one of the oldest sciences in the world, with recorded examples of botanical research and exploration dating back thousands of years. Humans interact with and use plants in a wide variety of ways, making it unsurprising that they have dedicated a great deal of energy to learning more about them, with the earliest attempts focused on finding plants which were safe to eat, while modern botany encompasses a range of activities.

There are a number of subfields within the field of botany, including plant taxonomy, plant pathology, phytoanatomy, plant genetics, phytochemistry, paleobotany, and ethnobotany. These fields of study range from research on plants which existed in earlier eras in the Earth's geologic history to investigations of plants which are used in traditional medicine, with the goal of learning how these plants work and how they might be applied to mainstream pharmaceuticals. Because plants play such an important role in human societies, botanists have a wealth of material to work with.

Dictionary of Botany is an authoritative and up-to-date reference book on all aspects of the study of plants. It offers concise and accessible explanations of terms from plant science, biogeology, evolution, earth history, and all the earth sciences, as well as up-to-date entries on more current fields of interest such as ecology, plant genetics, plant physiology, biochemistry, and cytology. In addition, the book offers good coverage of taxonomic groups and takes full account of recent taxonomical revisions. This book be will be a perfect reference tool for amateur botanists, students of botany and anyone interested in the study of plants.

Editor

A

A : Prefix meaning 'not' or 'without'.

Ab : Prefix meaning 'away from' or 'departing'.

Abaxial : Facing away from axis or stem, such as the lower surface of a leaf.

Abciss : Detach, separate from main structure.

Aberrant : Departing from the normal or usual.

Abiotic Factors : Non-living factors that can affect life, like soil, nutrients, climate, wind etc.

Abortive : Imperfectly developed; defective; barron.

Abscise : To cut off; hence abscission, abscissive, abscissile.

Abscisic Acid (ABA) : A growth inhibiting hormone enabling perennial plants to tolerate stressful conditions by promoting dormancy, stomatal closure and inhibiting growth.

Abscission : Abscission is a process of shedding or separating part of an organism from the rest of it. Common examples are that of, plant parts like leaves, fruits, flowers and bark being separated from the plant.

Abscission Zone : The zone at the base of the flower (pedicel), fruit (peduncle) or leaf (petioles), at which plant cells fray off, thereby facilitating the easy fall of these plant parts.

Absorption Field : An organised system of meticulously constructed narrow trenches, which are partially filled with washed gravel or crushed stone, into which a pipe is placed. Discharges from septic tanks are passed through these trenches.

Absorption Spectrum : Graph indicating the relative abilities of pigments to absorb various wavelengths of light.

Acaulescent : Lacking a distinct stem.

Accessory : Subsidiary, like a non-chlorophyll pigment.

Accessory Bud : A bud in addition to the main axillary bud, may be collateral or superposed.

Accidental : Accidental refers to the occurrences or existence of all those species that would not be found in a particular region under normal circumstances.

Acclimation : Acclimation refers to the morphological and/or physiological changes experienced by various organisms to adapt or accustom themselves to a new climate or environment.

Accrescent : Expanding after flowering, increasing in size with age, e.g. the calyx expanding around the base of a fruit.

Accumbent : A term referring to seeds in which the embryonic root is wrapped around and lies along the edges of the two cotyledons.

Acerose : Narrow with a sharp, stiff point.

Acetogenic Bacterium : An aerobic, gram negative bacteria, that is rod-shaped, which is made of non-sporogenous organisms that produce acetic acid as a waste product.

Acetyl CoA : Developed as an intermediate of carbohydrate/ fat/ protein oxidation in the citric acid cycle, Acetyle CoA is the aceylated form of coenzyme A.

Acetylene Block Assay : Determines the release of nitrous oxide gas from acetylene treated soil, which is used to estimate denitrification.

Acetylene Reduction Assay : This is used to estimate nitrogenase activity by measuring the rate of reduction of ethylene to acetylene.

Achene : A dry indehiscent 1-seeded fruit, from an either superior or inferior ovary of 1 carpel, with the seed not fused to the fruit wall; e.g. as in Ranunculaceae (from a superior ovary) and Asteraceae (from an inferior ovary and usually topped by the pappus and sometimes called a cypsela).

Acicular : Needle-shaped, as applied to some kinds of foliage.

Aciculate : Finely striated by minute, needle-like bristles.

Acid Soil : Type of soil common to areas with sandy soil, an abundance of organic matter, and heavy rainfall. Overlay acidic soil is harmful to plants, but slightly acidic soil can be beneficial to plants.

Acidic : Description of a substance that has a high concentration of hydrogen ions.

Acidophile : An organism that grows well in an acidic medium (up to a pH of 1).

Acorn : The fruit of an oak, not including the cup or peduncle. The nut of an oak.

Acro : Prefix meaning 'of or towards the tine'.

Acropetal : Produced in succession towards the apex.

Acrophyll : An upper leaf or frond, especially of high-climbing ferns.

Acroscopic : The side of the organ directed towards the apex of the axis on which it is bourn.

Acrostichoid : Acrostichum–like, of exindusiate sori densely covering the lower surface of the frond, or large areas of it; the lamina may be contracted or not.

Actino : Prefix meaning star-shaped or radial.

Actinomorphic : Of a flower with the parts in each whorl particularly sepals, petals and stamens not differing in shape, size or placement. The flower therefore can be bisected symmetrically in several planes.

Actinomycete : These are Gram positive, nonmotile, nonsporing, noncapsulated filaments that break into bacillary and coccoid elements. They resemble fungi, and most are free living, particularly in soil.

Actinorhizae : The association present between actinomycetous and roots of plants.

Activated Sludge : Sludge particles which are produced in raw or settled waste–water, by the growth of organisms in aeration tanks. This is all done in the presence of dissolved oxygen. This sludge contains living organisms that can feed on incoming waste-water.

Activation Energy : The amount of energy required to bring all molecules in one mole of a substance, to their reactive state, at a given temperature.

Active Carrier : An infected person who has visible clinical symptoms of a disease, and is capable of transmitting the disease to other individuals.

Active Site : The location on the surface of the enzyme where the substrate binds.

Active Transport : The forceful movement of molecules from one side of the plasma membrane to the other side against a diffusion gradient, by expenditure of energy.

Activity Space : The entire range of climatic and environmental conditions suitable to normal functions, process and activities of a living organism.

Aculeate : Bearing a sharp point.

Acuminate : Tapering gradually to a pointed apex with more or less concave sides along the tip.

Acute : Pointed, having a short sharp apex, the converging edges forming an angle of less than 900.

Acyclic : With the floral parts arranged spirally rather than in whorls.

Adaptation : Adaptation refers to the genetic mechanism of an organism to survive, thrive and reproduce by constantly enhancing itself, by altering its structure or function, in order to become better suited to the changing environment.

Adaptive Radiation : Diversification of group of organisms into several new species in order to fit into new environments.

Adaxial : Situated toward the axis, often on the upper side of branch.

Additive Genetic Effects **:** When the combined effects of alleles at different loci are equal to the sum of their individual effects.

Adenine (A) **:** A nitrogenous base, one member of the base pair AT (adenine-thymine).

Adenosine Diphoshate (ADP) **:** A nucleotide comprising adenine, two phosphate units and ribose, it is a cofactor contributing either phosphate group or energy or both to a reaction.

Adenosine Triphoshate (ATP) **:** A nucleotide comprising adenine, ribose and three phosphate units, is the major energy currency of the cell. It is a cofactor contributing phosphate group or energy or both to the reaction.

Adherent **:** Two or more organs appearing to be fused but actually separable.

Adhesion **:** Where two dissimilar parts or organs stick together but without organic fusion. adj. adherent.

Adhesion Hence Adherent **:** Of two dissimilar organs or parts touching each other, +/- adhesively but easily separated and not fused or grown together.

Adhesive Force **:** It is the force of attraction between dissimilar molecules due to which they stay together. Eg. Water droplets on a leaf.

Adiaxial **:** On the side of a lateral organ towards the axis or stem.

Adjuvant **:** The material added to an antigen to increase its immunogenicity, for example, alum.

Adnate **:** (1) fusion of unlike parts, e.g. stamens fused to the corolla, (2) of an anther which has a broad point of attachment by which it is rigidly held at the apex of the filament, as in some eucalypts (family Myrtaceae).

Adpressed **:** Closely pressed together but not united.

Adventitious **:** Occurring in unusual or unexpected locations such as roots on aerial stems or buds on leaves. Also meaning : out of the usual place, introduced but not yet naturalised.

Adventitious Bud **:** A bud produced anywhere on the wood surface, especially where the wood has been injured.

Adventitious Root : A general term used to describe any root that grows in an unusual orientation or location; examples include above ground roots.

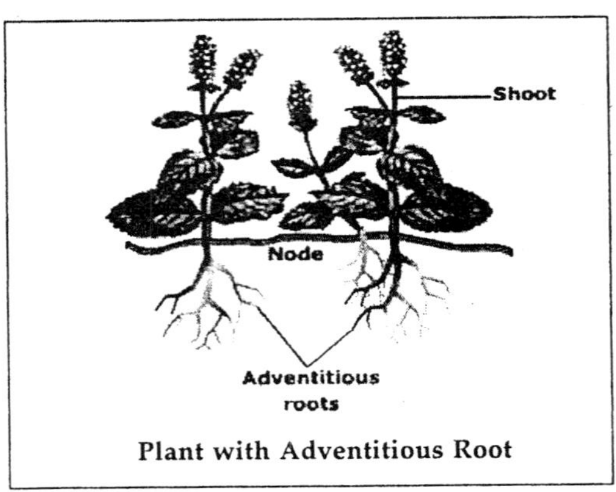

Plant with Adventitious Root

Adventive : Introduced accidentally, as most exotic weeds are; often used of introductions that are not fully naturalised.

Aequilateral : Equal-sided, as opposed to oblique.

Aerenchyma : A tissue of thin-walled cells with many air spaces, especially common in aquatic plants.

Aerial Behaviour : Aerial behaviour is a type of behaviour that deals with communicative or playful behaviour. It is most seen in whales and dolphins when they surface above water to either jump, leap, or just flit across.

Aerial Root : An adventitious root growing from the stem above ground level.

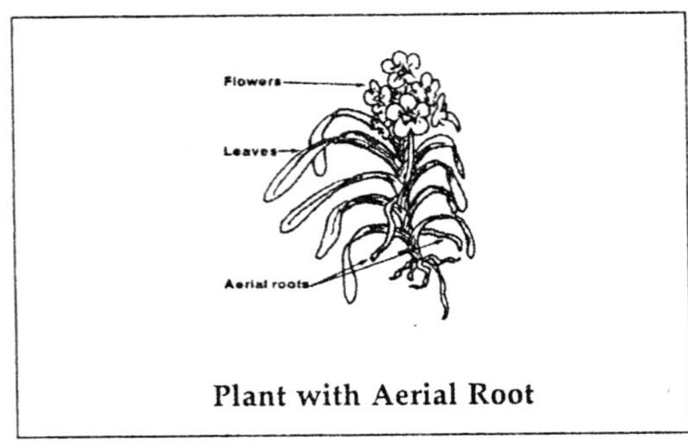

Plant with Aerial Root

Aerial Rootlets : Small hair like roots located on the stalk of some vines, such as ivy, adapted to clinging for support.

Aero : Prefix to do with 'air'; hence aeration etc.

Aerobic : This includes organisms that require molecular oxygen to survive (aerobic organisms), an environment that has molecular oxygen, and processes that happen only in the presence of oxygen (aerobic respiration).

Aerobic Anoxygenic Photosynthesis : Photosynthetic process which takes place under aerobic conditions, but which does not result in the formation of oxygen.

Aerobic Cellular Respiration : Part of cellular respiration, and plays a significant role in producing energy required to carry out different functions of the plant. It requires oxygen for the process.

Aerobic Respiration : Type of respiration requiring free oxygen as the terminal electron acceptor.

Aerophore : An outgrowth of tissue concerned with gas exchange, often at the base of pinnae in ferns.

Aerotolerant Anaerobes : Microbes that can survive in both, aerobic and anaerobic conditions, because they obtain their energy by fermentation.

Aestivate : To become dormant in summer.

Aestivation : Arrangement of the sepals and petals or their lobes in the unopened bud.

Affected Relative Pair : Individuals related by blood, each of whom is affected with the same trait. Examples are affected sibling, cousin, and avuncular pairs.

Aflatoxin : A toxin produced by Aspergillus flavus and Aspergillus parasiticus, which contaminate groundnut seedlings. This is said to be a cause of hepatic carcinoma.

Agamospermy : Asexual reproduction methods involving cells of only the ovule to yield seeds and fruit.

Agar : A dried hydrophilic, colloidal substance extracted from red algae species, used as a solid culture media for bacteria and other microorganisms. Also used as a bulk laxative, in making emulsions and as a supporting medium for immunodiffusion and immuno-electrophoresis.

Agarose : Agarose is obtained from seaweed and is used as a resolving medium in electrophoresis. It consists of non-sulfated linear polymer, which contains D-galactose and 3:6-anhydro-L-galactose alternately.

Agglutinates : The visible clumps that are formed as a result of an agglutination reaction.

Agglutination Reaction : The process of clumping together, in suspension of antigen bearing cells, microorganisms, or particles in the presence of specific antibodies called agglutinins. This leads to the formation of an insoluble immune complex.

Aggregate Fruit : Cluster of fruits derived from a single flower in which the carpels are free, or almost so, from each other. e.g. as in many Ranunculaceae, Annonaceae, Rosaceae.

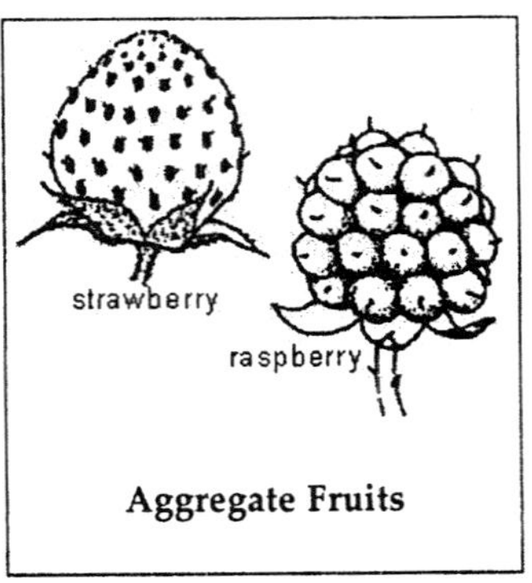

Aggregate Fruits

Aggregation Technique : A technique used in model organism studies in which embryos at the 8-cell stage of development are pushed together to yield a single embryo (used as an alternative to micro–injection).

Aigrette : A tuft of long and loose feathers used by breeding herons and egrets, during courtship displays.

Air Layering : A type of layering used on branches of tall plants. A slanting cut in the bark is made below the node. Rooting hormone is applied to the cut. Damp sphagnum moss is packed into the cut and then plastic is wrapped around to secure the moss and protect the cut from the elements. Roots will form in the moss after a few months after which the branch can be cut and planted in its own pot.

Airborne Transmission : A type of transmission, wherein the organism is suspended in or spreads its infection by air.

Akaryotic : Without a true nucleus.

Akinete : Enlarged cell with food reserves and thick cell wall which may undergo dormancy, e.g. in Cyano–bacteria.

Ala : A wing; hence alate winged or with wing-like appendages.

Alate : Having wings or wing-like structures.

Albuminous Seed : Seed containing large amounts of endosperm.

Alcoholic Fermentation : A fermentation process that produces alcohol (ethanol) and carbon dioxide from sugars.

Alga : Phototrophic eukaryotic microorganisms, that maybe unicellular or multicellular. These include phaeophyta: brown algae, spirogyra and red algae.

Alginate : Salt form of alginic acid, a polysaccharide colloid produced in walls of Phaeophyceae made of mannuronic acid and guluronic acid units.

Aliphatic : Pertaining to any member of one of the two major groups of organic compounds, with the main carbon structure as a straight chain.

Alkaline : Soils that contain high amounts of various salts of potassium and/or sodium, as well as other soluble minerals, and are basic rather than acidic with a Ph greater than 7.0

Alkaline Soil : A type of soil that commonly occurs in regions with light rainfall and has high levels of calcium carbonate. Overlay alkaline soil is harmful to plants, but slightly alkaline soil can be beneficial to plants.

Alkalophile : Organisms that have an affinity for alkaline media, thus, growing best in such conditions.

Allele : Alternative form of a genetic locus; a single allele for each locus is inherited from each parent (e.g., at a locus for eye colour the allele might result in blue or brown eyes).

Allelopathy : A characteristic of some plants according to which chemical compounds are produced that inhibit the growth of other plants in the immediate vicinity.

Alligator : A broad snouted crocodilians of the genus Alligator found in subtropical regions. This reptile is known for its sharp teeth and powerful jaws.

Allochthonous Flora : Organisms that are not originally found in soil, but reach there by precipitation, sewage, diseased tissue and other such means. They do not contribute much ecologically.

Allogeneic : Variation in alleles among members of the same species.

Alloparasite : Parasite not related to its host.

Allopatric : Organisms that occur, originate or occupy in separate geographical areas.

Allopatric Speciation : Speciation emerging as the result of physical separation of two or more population of one species, such that interbreeding is not possible.

Allosteric Site : A non-active site on the enzyme body, where a non-substrate compound binds. This may result in conformational changes at the active site.

Allotype : Any of various allelic variants of a protein, characterised by antigenic differences.

Alpha Diversity : A measurement of species richness in a natural unit (specified area) consisting of all plants, animals and microorganisms in a habitat functioning together.

Alpha Hemolysis : A partial clearing zone, greenish in colour, around a bacterial colony that grows on blood agar.

Alpha-proteobacteria : One of the five subgroups of proteobacteria, each with distinctive 16S rRNA sequences. Mostly contains oligotrophic proteobacteria, many of which have distinctive morphological features.

Alpine : Pertaining to or occurring on very high and cold mountains.

Alternate : Said of leaves which are not opposite each other on the axis but are borne singly at regular intervals at different nodes; of flower parts when they are situated at intervals between other parts, e.g., petals may be alternate with the sepals.

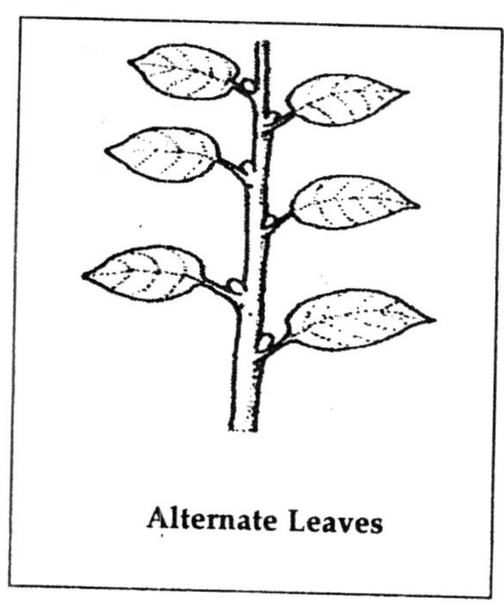

Alternate Leaves

Alternation of Generations : Plant life cycle type in sexually reproducing organisms involving alternation of diploid sporophyte phase and haploid gametophyte phase.

Alternative Complement Pathway : A pathway of complement activation, including the C3-C9 components of the classical pathway. It is independent of antibody activity.

Alternative Splicing : Different ways of combining a gene's exons to make variants of the complete protein.

Altruism : Instinctive behaviour performed towards the welfare of others, sometimes at personal cost.

Alveolar Macrophage : A highly active and aggressive phagocytic macrophage, located on the epithelial lining of the lung alveoli, which ingests and destroys any inhaled particles and microorganisms.

Alveolus : A small angular cavity, sac or pit in the body.

Ambulacra : This term refers to echinoderm's five part radial areas (under surfaced side) from where the tube feet protrude as well as withdraw.

Amensalism (Antagonism) : A type of symbiosis, wherein one population is adversely affected, while the other is unaffected.

Ament : A catkin; a dry scaly pendant spike of unisexual flowers, each subtended by a bract such as the inflorescence of birch or willow, usually deciduous in one piece. A tassel-like group of flowers.

Ames Test : A test that uses a special strain of *salmonella* to test chemicals for mutagenicity and carcinogenicity.

Amino Acid : Any of a class of 20 molecules that are combined to form proteins in living things. The sequence of amino acids in a protein and hence protein function are determined by the genetic code.

Amino Acid Activation : The first stage of synthesis of proteins, where the amino acid is attached to transfer RNA.

Amino Group : The monovalent radical NH_2, attached to a carbon skeleton, as seen in amines and amino acids.

Aminoacyl or Acceptor Site (A Site) : The site on the ribosome that contains an aminoacyl-tRNA at the beginning of the elongation cycle during protein synthesis.

Ammonia Oxidation : A test which is conducted during manufacturing process, to evaluate ammonia oxidation rate for nitrifiers.

Ammonification : Liberation of ammonia by microorganisms acting on organic nitrogenous compounds.

Amoeba : A minute protozoan, occurring as a single cell with a nucleus, that changes shape by extruding its cytoplasm, leading to the formation of pseudopodia, by means of which it absorbs food and moves.

Amoeboid Movement **:** Movement by means of extrusions of the cytoplasm, leading to formation of foot-like processes called pseudopodia.

Amphitrichous **:** A cell which has a single flagellum at each end.

Amphotericin B **:** An antibiotic derived from *streptomyces nodosus* which is effective against many species of fungi and certain species of *leishmania.*

Amplexicaul **:** Describing a sessile leaf that has its base completely surrounding the stem.

Amplexus **:** Mating position of the frogs and toads, in which the female sheds the eggs into the water and the male fertilises it. Fertilisation takes place outside of the female's body.

Amplification **:** An increase in the number of copies of a specific DNA fragment; can be in vivo or in vitro.

Amyloplast **:** Colourless organelle related to starch production.

Anabolism **:** Process of metabolism by which various small molecules are combined to form large ones.

Anadromic, Anadromous **:** With the first sub–branch of a lateral branch produced on the acroscopic margin, mostly of venation in bipinnate ferns.

Anaerobic **:** Refers to organisms that survive in the absence of oxygen (anearobic organisms), the absence of molecular oxygen, processes occurring in the absence of oxygen like anearobic respiration.

Anaerobic Respiration **:** Also called fermentation, this type of respiration does not need oxygen as the terminal electron acceptor.

Anamorph **:** A stage of fungal reproduction, where cells are asexually formed by the process of mitosis.

Anaplerotic Reactions **:** Reactions that help replenish intermediates in the tricarboxylic acid cycle when their reserves are depleted.

Anapsid **:** An extinct subclass of reptiles except for the **turtles,** that have no opening in the temporal region of the **skull.**

Anastomose : To join together, principally of veins.

Anastomosing : Fusing to form a network, as in the veins of a leaf.

Anastomosis : A network of intersecting or connecting blood vessels, nerves, or leaf veins that form a plexus.

Ancipital : Two-edged, such as the winged stem of *Sisyrinchium*.

Androecium : A collective term for the stamens of a flower.

Androgynophore : A stalk bearing both the stamens and superior ovary, e.g. in Passifloraceae.

Androgynous : Having staminate and pistillate flowers in the same inflorescence.

Anemophily : Seed plants which are pollinated by wind are said to be anemophilous.

Anergy : Decreased responsiveness to antigens, to the extent that there is an inability to react to substances that are expected to be antigenic.

Aneuploid : Anomaly in the usual chromosome number, wherein one or more chromosomes are missing or present as extras.

Angiosperm : n. A group of plants that produce seeds enclosed within an ovary, which may mature into a fruit; flowering plants.

Angiosperms : The flowering plants; plants with ovules enclosed in an ovary.

Angled : Sided, as in the shape of stems or fruits.

Angular : Having sharp angles or corners, generally used in reference to structures such as stems to contrast them with rounded stems.

Anion Exchange Capacity : Total exchangeable anions that a soil can adsorb. The unit used to express the amount is in centimoles of negative charge per kilogram of soil.

Anisogamy : Reproduction by motile gametes that differ in morphology or behaviour.

Anisophyllous : Of leaves, usually a pair, of differing size and shape.

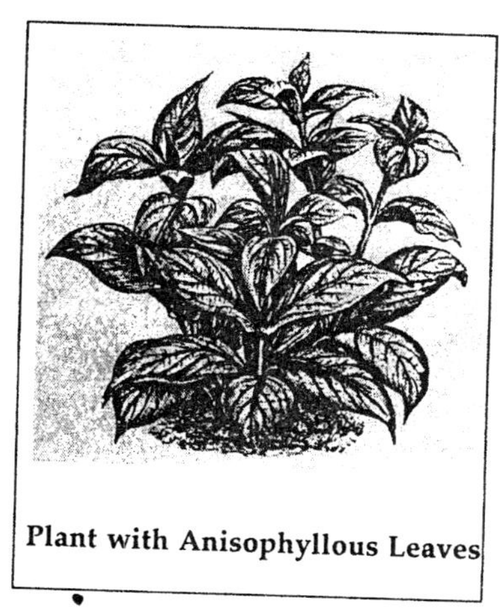

Plant with Anisophyllous Leaves

Annotation : Adding pertinent information such as gene coded for, amino acid sequence, or other commentary to the database entry of raw sequence of DNA bases.

Annual : Plant which completes its life cycle, from germination, to flowering, to seed production, in a single year or less. Examples include grains, beans, and many wild flowers.

Annual Ring : The formation of wood in plants on an annual basis comprises two concentric layers of wood : spring wood and summer wood.

Annular : In the form of a ring (n. sing. annulas).

Annulus : The arrangement of thick-walled cells involved in opening the sporangium in ferns.

Anoxic : A condition or state which is devoid of oxygen.

Anoxygenic Photosynthesis : A type of photosynthesis where oxygen is not produced. This phenomenon is seen in green and purple bacteria.

Antagonist : A drug that binds to a hormone, neurotransmitter, or another drug, thus, blocking the action of the other substance.

Anterior : Away from the axis, toward the subtending (enclosing) bract.

Anther : Male reproductive part of an angiosperm. Specifically, the terminal pollen sac of the stamen where male gametes in the form of pollen grains are produced.

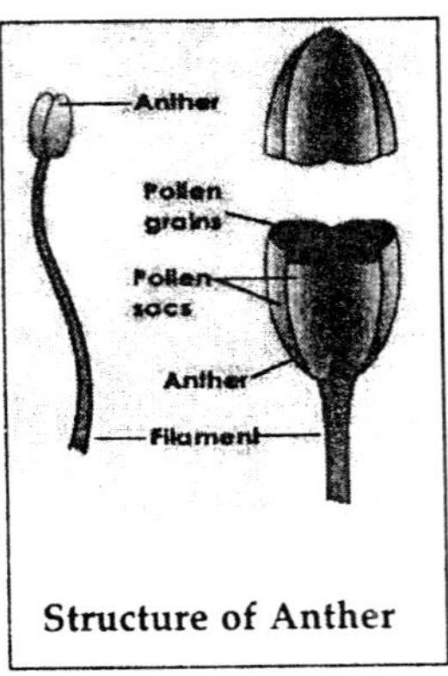

Structure of Anther

Antheridium : The male gametangium found in phylum Oomycota (kingdom Stramenopila) and phylum Ascomyta (kingdom Fungi).

Anthesis : The time in the development of a flower when it is opening.

Anthocarp : A false fruit consisting of the true fruit surrounded by the base of the perianth, as in Nyctaginaceae.

Anthocyanin : Water soluble pigment located in the cell sap, which varies from red to blue in colour. Found in flowers of most plants.

Anthophyte : A flowering plant, or any of its closest relatives, such as the Bennettitales, Gnetales, or Pentoxylales.

Anthotelic (Determinate Inflorescence) : An inflorescence with the inflorescence or parts of the inflorescence ending in a flower or an aborted but distinctly floral bud, e.g. panicle, thyrsoid, dichasium, monochasium.

Anthropogenic : Something that is derived from human activities.

Antibiosis : Lysis of an organism brought about by metabolic products of the antagonist. This can be caused by enzymes, lytic agents or other toxic compounds.

Antibiotic : A chemical substance produced by a microorganism, which has the capacity to inhibit the growth of, or kill other microorganisms.

Antibody : An immunoglobulin molecule that reacts with a specific antigen that induced its synthesis and with molecules that have a similar structure.

Antibody-Dependent Cell-Mediated Cytotoxicity (ADCC) : A type of reaction wherein, cells with Fc receptors that recognise the Fc region of the bound antibody, kill the antibody-coated target cells.

Anticipation : Each generation of offspring has increased severity of a genetic disorder; e.g., a grandchild may have earlier onset and more severe symptoms than the parent, who had earlier onset than the grandparent.

Anticodon Triplet : A triplet of nucleotides in transfer RNA that is complementary to the codon in messenger RNA.

Antigen : Any substance capable of instigating the immune system into action, inciting a specific immune response and of reacting with the products of that response.

Antimetabolite : A substance that interferes with a specific metabolic pathway, by inhibiting a key enzyme, due its resemblance with the normal enzyme substrate.

Antimicrobial Agent : An agent that has the capacity to kill or inhibit the growth of microorganisms.

Antisense : Nucleic acid that has a sequence exactly opposite to an mRNA molecule made by the body; binds to the mRNA molecule to prevent a protein from being made.

Antisense RNA : One of the strands of a double-stranded molecule, which does not directly encode the product, but is complementary to it, thus, inhibiting its activity.

Antiseptic : A substance that inhibits the growth and development of microorganisms, but does not necessarily kill them.

Antler : One pair of bony, deciduous and branched hornlike structure found on the head of a deer, moose, elk, etc.

Apex : The tip of a plant part.

Apical : Of the apex or attached at the apex or top, e.g. ovules attached to an apical placenta'.

Apical Deristem : Meristem located at the tip of the root, shoot or other organs of the plant.

Apical Dominance : A phenomenon whereby a plant's growth is concentrated on the terminal bud, allowing it to grow taller, thereby increasing its exposure to sunlight.

Apical Meristem : Meristems located at the terminal shoot of a plant, the points at which upward growth happens.

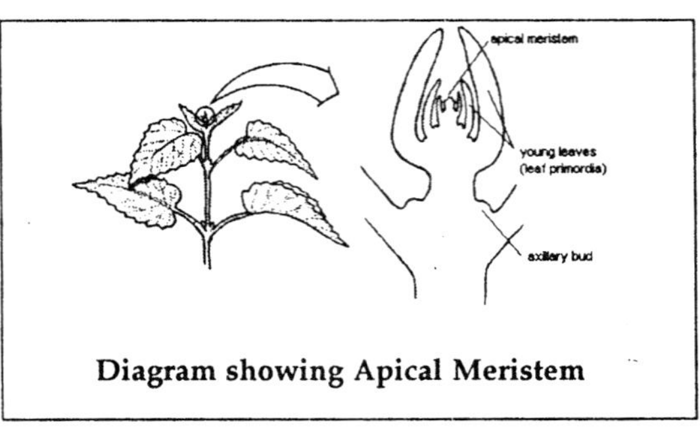

Diagram showing Apical Meristem

Apiculate : With a small abrupt point which is demarcated from the organ to which it is attached, e.g. of some anthers.

Apiculus : A short sharp point in which an organ may end; hence apiculate.

Aplanospore : A spore that is formed during asexual reproduction, which is nonflagellated and nonmotile.

Apocarpous : A gynoecium consisting of two or more carpels which are free and distinct from each other, e.g. as in Ranunculaceae and Dilleniaceae.

Apoenzyme : A protein part of an enzyme that is separable from the prosthetic group (the co-enzyme).

Apogamy : The development of the sporophyte from the prothallus without fertilisation.

Apomeiosis : Nuclear division without meiosis.

Apomixis : The process whereby a plant produces viable seed without fertilisation. adj. apomictic.

Apomorph : A new specialised trait in an evolving organism which is completely different from its ancestral line.

Apoplast : A route of transport within a plant consisting of the extra cellular space made up of cell walls.

Apoptosis : A pattern of cell death which is often called 'programmed death' or 'suicide of cells', wherein the cell breaks up into fragments, which are membrane bound. These fragments are then eliminated by phagocytosis. This is a protective mechanism, by which the cell prevents spread of infection to other cells by sacrificing itself.

Aporepressor : A product of regulator genes, that combines with the corepressor to form the complete repressor.

Apospory : The development of prothalli from direct outgrowth the fern frond, without the production of spores.

Appendage : An attachment developed on and projecting beyond the surface of an organ.

Applanate : Flattened out and horizontally expanded.

Appressed : Closely applied to the supporting organ or axis for the entire length (also adpressed).

Arachnoid : Cobweb-like; formed of tangled hairs or fibres.

Arboreal : Arboreal refers to animals that have adapted themselves to live and move in the trees.

Arborescent : Approaching the size and habit of a tree.

Arbuscule : Special structure formed in the root cortical cells by arbuscular mycorrhizal fungi. The structure formed resembles a tree.

Archegonium : The organ on a gametophyte plant which produces the egg cell, and nurtures the young sporophyte.

Arcuate : Curved or arched, fairly strongly.

Areolate : Of surface pattern or venation, divided into many angular or squarish spaces, e.g. the venation and surface pattern in dried specimens in many Lauraceae.

Areole : (1) in Cactaceae, a cluster of hairs and/or spines borne at the node of a leafless stem; (2) a space in any reticulated surface, e.g. space between veins. adj. areolate.

Aril : An expansion of the funicle into a fleshy or membranous appendage, sometimes partially or wholly covering the surface of the seed, and often brightly coloured, as in some Sapindaceae.

Arista : An awn or bristle; hence aristate, bearing or tapering into an awn or bristle (pl. aristae; dimin. aristulate).

Aristate : Having a stiff, bristle-like awn or tip.

Armature : Covering or occurrence of spines, hooks or prickles; hence armed.

Armed : Provided with spines, tubercles or comparable structures.

Aromatic : With a resinous, spicy or distinctive smell.

Arrayed Library : Individual primary recombinant clones (hosted in phage, cosmid, YAC, or other vector) that are placed in two-dimensional arrays in microtiter dishes. Each primary clone can be identified by the identity of the plate and the clone location (row and column) on that plate. Arrayed libraries of clones can be used for many applications, including screening for a specific gene or genomic region of interest.

Article : (1) part of an organ which separates readily from the rest of an organ, e.g. as in a lomentum; (2) portion of branchlet between whorls of teeth in Casuarinaceae.

Articulate : With one or more joints or points of apparent separation, usually marked by a swelling, line or abrupt change in colour.

Articulated : Jointed; usually separating at the point of articulation into segments or article, e.g. *Calliarthron* or *Halimeda*.

Artificial Selection : A selection process where the breeder chooses the animals for mating and producing offspring's of desired inheritable qualities.

Artificially Acquired Passive Immunity : A type of temporary immunity that results from the introduction of antibodies produced by another organism or by in vitro methods, into the body.

Ascending : Growing obliquely at first but finally upwards; rising or growing upwards.

Aseptic Technique : Procedures that are performed under strict sterile conditions. These procedures maybe laboratory procedures such as microbiological cultures.

Asexual : Without the involvement of ferilisation; of propagation by division or the production of bulbils or stolons etc.

Asexual Reproduction : Reproduction by one sex via budding, fission, etc. resultant in genetically identical offspring.

Aspect Diversity : It is the measure of the different physical appearances that are found in a group of species living in a common habitat and are hunted by other animals that use visual hunting skill to identify and kill their prey.

Aspera : Rough to the touch; hence asperites, aspirate; rough or harsh (dimin. asperulous).

Assembly : Putting sequenced fragments of DNA into their correct chromosomal positions.

Assimilatory Nitrate Reduction : Reduction of nitrate to compounds like ammonium, for the synthesis of amino acids and proteins.

Associative Dinitrogen Fixation : An enhanced rate of dinitrogen fixation, brought about by a close relationship between free-living diazotrophic organisms and a higher plant.

Associative Symbiosis : Interaction between two dissimilar organisms or biological systems, which is normally mutually beneficial.

Asymmetric : Of a leaf, leaf base or other organ, having the sides unequal.

Atom : A unit of matter consisting of electrons, protons, neutrons, and other subatomic particles.

Atomic Number : The number of protons within an atom's nucleus. The number of protons in a normal atom is always equal to the number of neutrons.

Atomic Weight : The total mass of an atom is comprised of the masses of electrons, protons, neutrons, and subatomic particles.

ATP : Adenosine Triphosphate, a molecule that when undergone a dephosphorylation reaction (a loss of a phosphate) releases energy which can be used by cells.

Atro : Prefix meaning 'dark'.

Attenuate : Gradually becoming very narrow or slender.

Auricle : (1) an ear-like outgrowth at the base of the sheath of some grasses and other monocots; (2) an ear-shaped lobe at the base of a leaf or other organ. adj. auriculate.

Auriculate : With earlike appendages; eared.

Autogenous Infection : An infection which occurs due to the microbiota of the patient himself.

Autoimmune Disease : A disease where the target is the body's own tissues, that is, there is attacking of self-antigens.

Autoimmunity : A condition where a specific humoral or cell mediated immune response is initiated against the constituents of the body's own tissues. It normally leads to hypersesitivity reactions, and if it persists, can even escalate to an autoimmune disease.

Autolysins : A lysin that originates in an organism, which is capable of destroying its own cells and tissues.

Autonym : The name of a species automatically applied at the infraspecific level to the type when another infrospecific taxon is described in that species.

Autoradiography : A technique that uses X-ray film to visualise radioactively labelled molecules or fragments of molecules; used in analysing length and number of DNA fragments after they are separated by gel electrophoresis.

Autosomal Dominant : A gene on one of the non-sex chromosomes that is always expressed, even if only one copy is present. The chance of passing the gene to offspring is 50% for each pregnancy.

Autosome : A chromosome not involved in sex determination. The diploid human genome consists of a total of 46 chromosomes : 22 pairs of autosomes, and 1 pair of sex chromosomes (the X and Y chromosomes).

Autotroph : Organisms that obtains food/energy by not eating other organisms. For example autotrophs use energy from the sun or obtain energy from inorganic substances. Plants are autotrophs, specifically photoautotrophs.

Autotrophic : Independent of other organisms in respect of organic nutrition, able to form carbohydrates by process of photosynthesis.

Autotrophic Nitrification : The combined nitrification action of two autotrophic organisms, one converting ammonium to nitrite and the other oxidising nitrite to nitrate.

Autotropic : Organisms converting inorganic matter into organic material for the purpose of sustenance.

Auxiliary Cell : Cell in post-fertilisation development in the Rhodophyta which receives the diploid (2N) zygote nucleus and then develops the gonimoblast filaments.

Auxins : Plant hormones that have a variety of effects on the cellular responses of plants. For example development of leaves and fruit, also formation of secondary growth.

Auxotroph : A mutated type of organism that requires specific organic growth factors, in addition to the carbon source present in a minimal medium.

Avuncular Relationship : The genetic relationship between nieces and nephews and their aunts and uncles.

Awn : A slender, stiff terminal bristle attached at its base to another structure or organ such as a leaf or grass stem.

Axenic : Pure cultures of microorganisms, that is, which are not contaminated by any foreign organisms.

Axial Filament : Found in spirochetes, it is the organ of motility.

Axil : The angle formed between a leaf stalk and the stem to which it is attached. In flowering plants, buds develop in the axils of leaves.

Axile : (1) on the axis; (2) of placentation, with the placentas and ovules along the central axis of the ovary in a compound ovary with septa.

Axillary : In or related to an axil, e.g., axillary buds; occurring in an axil.

Axillary Bud : An embryonic side shoot. A point on a stem, at the node, and between the stem and leaf, where a new shoot can develop. Growth is usually inhibited at these buds. Axillary buds have the potential to give rise to either shoots bearing flowers, or to a vegetative branch which includes its own terminal bud.

Axis : Main stem or central line of development on which secondary or side branches are borne.

B

Baccate : Like a berry, having berries.

Bacillus : Rod shaped, spore-producing bacteria belonging to the genus Bacillus.

Backflow Preventer : A device used in irrigation systems to prevent back flow of water.

Bacteremia : Presence of bacteria in the blood.

Bacteria : Single celled, omnipresent organisms appearing in spiral, spherical or rod shape.

Bacterial Artificial Chromosome (BAC) : A vector used to clone DNA fragments (100- to 300-kb insert size; average, 150 kb) in *Escherichia coli* cells. Based on naturally occurring F-factor plasmid found in the bacterium *E. coli.*

Bacterial Artificial Chromosome : A cloning vector that is derived from E. coli, which is used to clone foreign DNA fragments in E. coli.

Bacterial Photosynthesis : A mode of metabolism, which is light-dependent and where carbon dioxide is reduced to glucose, which is used for energy production and biosynthesis. It is an anaerobic reaction.

Bactericide : A substance that kills bacteria.

Bacteriochlorophyll : A light absorbing pigment found in phototrophic bacteria, like green sulfur and purple sulfur bacteria.

Bacteriocin : Substances that are produced by bacteria which kill other strains of bacteria by inducing a metabolic block.

Bacteriophage : It is an obligate intracellular parasite that breed inside bacteria by using the host's cellular machinery.

Bacteriorhodopsin : A protein involved in light mediated ATP synthesis, which contains retinal. It is one of the main characteristics of archaebacteria.

Bacteriostatic : An agent that inhibits the growth or multiplication of bacteria, but does not kill them.

Bacteroid : A genus of *bacteroides*, these are Gram negative, rod-shaped, anaerobic bacteria which are normal inhabitants of the oral, respiratory, urogenital and intestinal cavities of animals and humans.

Baeocytes : Reproductive cells formed by cyanobacteria through multiple fission. They are small and spherical in shape.

Balanced Growth : Microbial growth where all cellular constituents are synthesised at constant rates, in relation to each other.

Balanced Polymorphism : A situation where more than one allele is maintained in a population, which is the outcome of the heterozygote being superior to both homozygotes.

Banner : The upper petal of a pea flower.

Barbate : Bearded, provided with tufts of long, weak hairs.

Barbed : Bearing sharp, spine-like hooks which are bent backwards (dimin. barbellate).

Bare Root : Refers to the exposure of roots characteristic of some plants packaged for sale. A common example is the rose.

Bark : Technically, the tissue comprised of phloem, phelloderm, cork cambium, and cork external of the vascular cambium. Bark occurs in plants that have secondary growth.

Barophile : An organism that thrives in conditions of high hydrostatic pressure.

Barotolerant : An organism that can tolerate high hydrostatic pressure, although it will grow better under normal pressure.

Barren : Sterile, incapable of reproducing.

Basal : (1) (radical) attached or grouped at the base, e.g. of leaves in a rosette (2) of placentation, with the placenta at the base of the ovary.

Basal Body : A cylindrical structure that attaches the flagella to the cell body at the base of prokaryotic or eukaryotic organisms.

Basal Medium : A basal medium allows the growth of many types of microorganisms which do not require special nutrient supplements.

Base : A substance that has a low concentration of hydrogen ion and conversely has a high concentration of hydroxyl ions.

Base Composition : The proportion of total bases consisting of guanine plus cytosine or thymine plus adenine base pairs.

Base Pair (bp) : Two nitrogenous bases (adenine and thymine or guanine and cytosine) held together by weak bonds. Two strands of DNA are held together in the shape of a double helix by the bonds between base pairs.

Base Sequence : The order of nucleotide bases in a DNA molecule; determines structure of proteins encoded by that DNA.

Base Sequence Analysis : A method, sometimes automated, for determining the base sequence.

Basi : Prefix meaning 'of or towards the base'.

Basic : Refers to a substance that has a high concentration of base. The pH is above 7.

Basidiocarp : Fruiting body in basidiomycete fungi, such as puffball or mushroom.

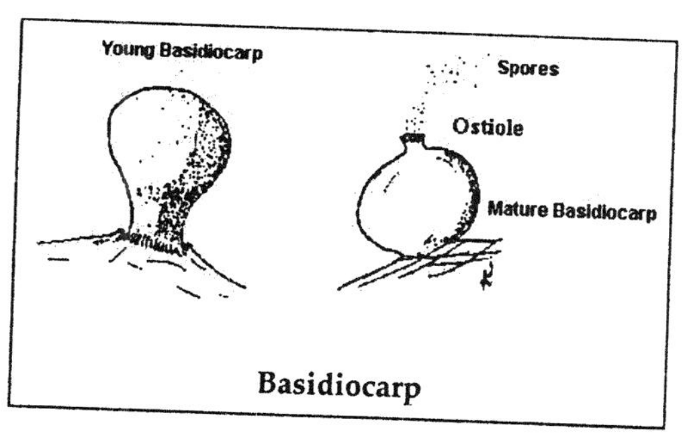

Basidiocarp

Basidioma **:** Fruiting body that produces the basidia.

Basidiospore **:** The sexual spore of the Basidiomycotina, which is formed on the basidium.

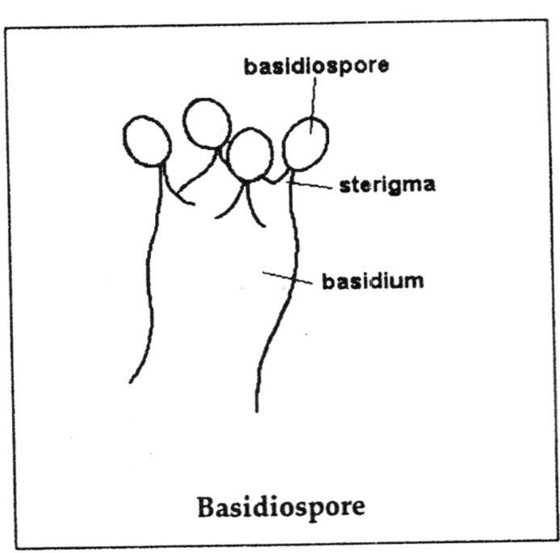

Basidiospore

Basidium **:** The cells in basidiomycete fungi in which fusion of nuclei and meiosis occur to produce basidiospores.

Basifixed **:** Attached at or by the base, e.g. of anthers attached by the base to the filament.

Basionym **:** The synonym (viz.) or combination (viz.) from which the specific epiphet was derived.

Basipetal **:** Produced in succession towards the base.

Basiscopic **:** The side of the organ directed towards the base of the axis on which it is borne.

Batch Culture **:** A culture of microorganisms which is obtained by inoculating a dish containing a single batch of medium.

Batch Process **:** A treatment procedure wherein, a tank or reactor is filled, the solution is treated, and the tank is emptied. Batch processes are mostly used to cleanse, stabilise, or condition chemical solutions for use in industries.

Batesian Mimicry **:** In a situation where a harmless species has evolved to replicate the warning signal given by a harmful species (directed at a common predator), Batesian mimicry occurs.

Bathyphyll : A lower leaf or frond, especially of high-climbing ferns.

B-cell (B Lymphocyte) : Bursa-dependent lymphocytes which are precursors of antibody-producing cells (plasma cells) and the cells primarily responsible for humoral immunity.

B-cell Antigen Receptor (BCR) : The membrane which is formed of membrane immunoglobulin or surface immunoglobulin, which allows a B-cell to detect, when a specific antigen is present in the body, and triggers B-cell activation.

Behavioural Genetics : The study of genes that may influence behaviour.

Benthic : A Benthic zone is the ecological region that encompasses the bottom most level of any body of water, be it a river, lake or ocean. When used in conjunction with a living organism, it refers to bottom-dwelling.

Benthic Zone : The ecological region at the lowest level of a water body, including the sediment surface and some subsurface layers.

Berry : A fleshy or pulpy indehiscent fruit with 1 or more seeds, the seeds embedded in the fleshy tissue of the pericarp; may be formed from either a superior or an inferior ovary.

Beta Diversity : A term of measurement, that gauges the variety of organisms in a region. It is impacted by the turnover of species among habitats.

Beta Hemolysis : A clear zone seen around a bacterial colony growing on blood agar.

Biaxial : Having two axial cell rows.

Bicolorous : With white or clear walls and dark red or brown septae as in some hairs.

Biconvex : In section with both surfaces curved outwards away from the centre.

Biennial : Plants requiring two seasons to complete their life cycle. The first season growth is purely vegetative and the second one bears fruit.

Bifarious : Arranged in two opposite rows.

Bifid (2-fid) : Divided into two parts, usually to about halfway.

Biflorous : Flowering in the spring and again in the autumn.

Bifoliolate (2-foliolate) : Of a compound leaf, with two leaflets.

Bifurcate : Divided into two forks or branches.

Bilateral : Of, on, or with equal sides, eg bilaterally symmetric, bilaterally flattened or compressed etc.

Bilateral Symmetry : This type of symmetry is exhibited by most animals, and just means that if a line were drawn down the middle of the body, both sides would be equal and symmetrical.

Bilocular (2-locular) : Having two cavities, e.g. of ovary or anther.

Binary Fission : Process of cell division in prokaryotes, such as yeasts where the cell devides into two daughter cells.

Binate : Almost or quite divided into two parts, eg leaflets, bristles etc.

Binomial Classification : System of classification that provides scientific names to organisms. Each name consists of a genus name and a species name.

Bioaccumulation : Intracellular accumulation of chemical substances in living tissue.

Bioaugmentation : Addition to the microorganisms environment that can metabolise and grow on specific organic compounds.

Bioavailability : The extent to which a drug or other substance becomes available to the target tissue after administration.

Biochemical Oxygen Demand : The amount of dissolved oxygen consumed in five days by biological processes breaking down organic matter. It is a test that measures the oxygen consumed (in mg/L) over five days at 20 degrees Celsius.

Biodegradable : The property by which a substance is capable of being degraded by biological processes, like bacterial or enzymatic action.

Biodegradation : The process of breakdown of substances by chemical reactions, thus rendering these substances less harmful to the environment.

Biodiversity : A term of measurement, that gauges the diversity of organisms in a habitat or ecosystem. This measurement can be made based on the number of species or genetic variation that exist within an ecosystem or region.

Biogeography : It is a term used to define the study of the geographic distribution of organisms throughout a region over a given period of time. It is carried out with the aim of examining where organisms dwell, and at what populations.

Bioinformatics : The science of managing and analysing biological data using advanced computing techniques. Especially important in analysing genomic research data.

Bioinsecticide : A pathogen (either bacteria, virus or fungi) used to kill or inhibit the activity of unwanted insect pests.

Biological Controls : Use of natural inhibitors or enemies to combat pests and other damage causing organisms.

Bioluminescence : The production of light in living organisms by the enzyme luciferase.

Biomagnification : Increase in the concentration of a chemical substance, as its position progresses in the food chain.

Biomass : Total mass of living matter present in a given habitat, expressed as volume of organisms per unit of habitat's volume or weight per unit area.

Biome : A region that is defined based on its climate and geography, which has ecologically similar communities of plants, animals, and soil organisms. The similarity is based on plant structures (such as trees, grasses and shrubs), plant spacing (forest, savanna, woodland), leaf types (such as needle leaf and broadleaf), and climate.

Bioremediation : The use of biological organisms such as plants or microbes to aid in removing hazardous substances from an area.

Biosphere : The portion of earth that is inhabited by life forms.

Biostimulation : A process which helps catalyse the activity of microorganisms involved in biodegradation.

Biosynthesis : Production of cellular constituents from simpler compounds.

Biota : They constitute the living components (flora and fauna) of an ecosystem, biome or habitat.

Biotechnology : A set of biological techniques developed through basic research and now applied to research and product development. In particular, biotechnology refers to the use by industry of recombinant DNA, cell fusion, and new bioprocessing techniques.

Biotic : Referring to things which have life.

Bio-Tower : A tower filled with a media similar to a rachet or plastic rings, where air and water are forced up the tower by a counter flow movement. It is an attached culture system.

Biotransformation : The chemical alterations of a drug, occurring in the body, due to enzymatic activity.

Biotrophic : Close associations seen between two different organisms, that work mutually to benefit each other.

Bioventing : A procedure where the subsurface is aerated to enhance biological activity of naturally occurring microorganisms in the soil.

Bipartitite : Divided into two +/- equal part, to the base or almost so.

Bipedal : Bipedalism is a manner of moving on land, where the organism progresses using only its two rear limbs, or legs.

Bipinnate (2-pinnate) : Of a compound leaf, with the lamina divided twice pinnately, i.e. with the pinnae themselves divided pinnately into pinnules.

Bipinnate : Describing a pinnate leaf in which the leaflets themselves are further subdivided in a pinnate fashion.

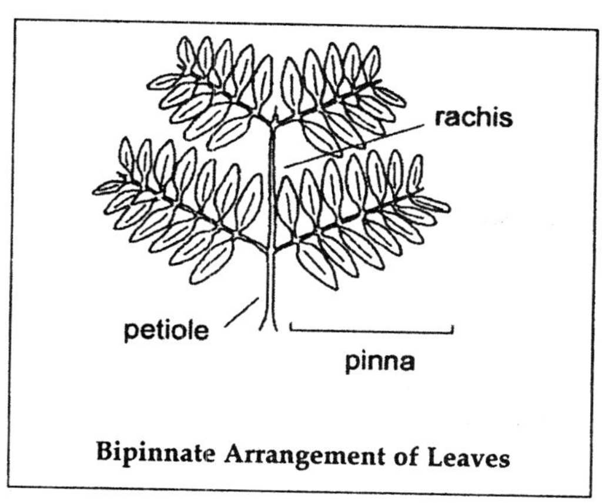

Bipinnate Arrangement of Leaves

Bipinnatifid (2-pinnatifid) : Of a simple leaf, with the primary lobes cut into smaller lobes (i.e. lobes pinnatifid).

Birth Defect : Any harmful trait, physical or biochemical, present at birth, whether a result of a genetic mutation or some other nongenetic factor.

Birth Rate : The term is the average number of young produced within a specific period of time. It is calculated per individual, and is usually communicated as a function of age.

Biseriate (2-seriate) : Arranged in two rows or whorls.

Bisexual : Of a flower, with both stamens and carpels present and functional.

Bisporangia : Sporangia that divides to produce 2 spores perhaps by mitotic division, in Corallinales and some other Rhodophyta.

Bisporangiate : When a flower or cone produces both megaspores and microspores, it is said to be bisporangiate. Most flowers are bisporangiate.

Biternate (2-ternate) : Twice ternate, the 3 pinnae each divided into 3 pinnules (a total of 9 pinnules).

Blade (lamina) : The expanded terminal portion of a leaf, petal or other structure, i.e. that portion of the leaf that does not include the stalk.

Blanching : The process of blocking light from certain types of vegetables in order to prevent coloration and keep mild flavours.

Blastomycosis : An infection caused due to Blastomyces dermatitidis, it predominantly affects skin, lungs and bones.

Blastotelic (indeterminate Inflorescence) : An inflorescence or part of an inflorescence not ending in a flower, i.e. ending in a non-floral bud, e.g. a thyrse, raceme or spike.

Bloom : A whitish powder which covers a usually waxy surface, but is easily rubbed off, e.g., on prune or plum.

Blowhole : A blowhole is an opening on the top of a cetacean's head, from which air is inhaled and exhaled.

Bole : The trunk of a tree below the lowest branch.

Bonsai : Literally meaning tray planting. The Japanese art of dwarfing plants.

Boss : A protuberance with a rounded surface.

Botryoid : A term describing an inflorescence of similar form to a botryum but ending in a flower or floral bud; includes raceme-like, spike-like, umbel-like and other variants.

Botryum : A simple inflorescence ending in a vegetative (non-floral) bud in which the main axis bears lateral flowers; includes racemes, spikes, umbels and corymbs.

Bow Riding : It is an activity carried out by cetaceans (most commonly dolphins), in which they swim or drift along the crests of waves in the ocean.

Brackish : A mixture of salt and fresh water, somewhat saline.

Bract : A modified leaf which may be reduced in size or different in other characteristics from the foliage leaves and which usually subtends a flower or an inflorescence.

Bracts : Leaves that grow just below a flower. Bracts are modified and usually green, but are sometimes colourful and mistaken for petals.

Branch : Any division or subdivision from the stem except the growth of the current season.

Branchlet : The ultimate divisions of a branch. As applied to woody plants, the growth of the season.

Breeding System : A breeding system includes all the different breeding behaviours (polygyny, outcrossing, or selective mating) of a population, and the methods in which the members of the population adapt to them.

Broad Leafed : Refers to plants that have foliage year round, but are not conifers. Also refers to any weed that is not a grass.

Bryophyte : Plants in which the gametophyte generation is the larger, persistent phase; they generally lack conducting tissues. Bryophytes include the Hepaticophyta (liverworts), Anthocerotophyta (hornworts), and Bryophyta (mosses).

Bud : An undeveloped or rudimentary stem or branch or flower, with or without scales.

Bud Union : The place at which the bud/shoot joins the root stalk.

Budding : Type of asexual reproduction involving formation of new cells from protrusions arising from mature cells. Yeast reproduces via budding.

Bulb : In general any structure underground from which a plant can grow. A true bulb is an underground structure that contains an embryonic plant protected by scales. An example of a true bulb is an onion.

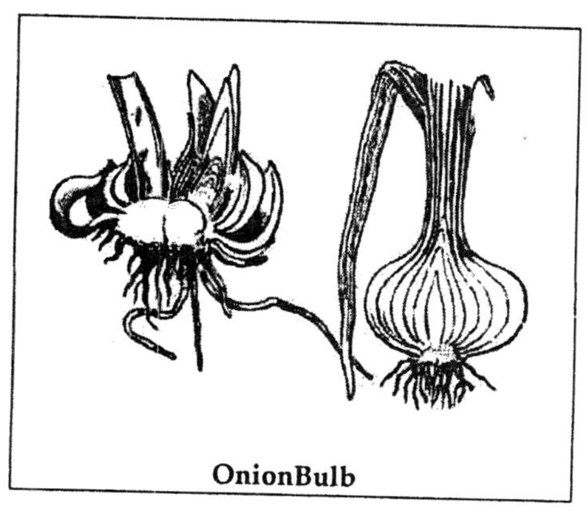

OnionBulb

Bulbiferous : Bearing bulbs or bulbils for vegetative propogation.

Bulbil : A small bulb formed in the axil of a leaf or bract and functioning to propagate the plant vegetatively.

Bulblet : A small bulb produced at the base of a bulb.

Bulbous : Bulb-shaped; swollen like an onion.

Bullate : Bubblelike; puckered or blistered, describing a layer of tissure (part of a leaf, scales etc) strongly arched and raised between the margins.

Bundle Sheath : Layer of parenchyma or sclerenchyma cells encircling the vascular bundle in plant leaves and stems.

Bundle-scar : A "dot" or mark on the surface of a leaf-scar, left by the breaking of the vascular bundles which extend through the petiole to the blade of the leaf.

Bur : A rough or spiny covering of a fruit.

Burl : A woody swelling where the stem joins the roots.

Burr : A prickly propagule consisting of a seed or fruit and associated floral parts.

Burst Size : The number of phages ejected by a host cell over the course of its lytic life cycle.

Butanediol Fermentation : A kind of fermentation found in Enterobacteriaceae family, where 2,3-Butanediol is a major product.

Buttress : A flange protruding from the lower part of the trunk, frequent in rainforest trees.

C

C3 Plant : Plants whereby the first organic product of carbon-fixation (during the Calivin cycle) is a three-carbon compound. This category includes the vast majority of common plant species.

C4 Plant : Plants whereby the first organic product of carbon-fixation is a four-carbon compound, which is formed prior to the Calvin cycle. This adaptation is advantageous in hot regions with intense sunlight as it minimises photorespiration and maximises sugar production. Sugarcane is an example of this type of plant.

Caching : Storing of food for later use, when food is not available or is short in supply.

Caducous : Falling off very early as compared to similar structures in other plants.

Caespitose : Tufted or matted, growing in tufts or patches.

Caiman : A tropical American crocodilian amphibian, found in Central and South America.

Calcereous : Calcium contained parts such as shells, bones and exoskeleton, which protects an animal, e.g. in the Corallinales.

Calcium : (Ca), a common element playing a fundamental role in cell formation and cell growth.

Call Matching : This is a behavioural trait, often displayed by members of the finch family. This refers to the male and female of a pair duplicating each others flight call, vocally.

Callose : A plant polysaccharide composed of glucose residues linked together through â-1, 3-linkages secreted by an enzyme complex (callose synthase), resulting in the hardening or thickening of plant cell walls.

Callosity : A thickened and hardened swelling on the surface of an organ.

Calvin Cycle : Biochemical reactions cycle occurring during photosynthesis in the chloroplasts, wherein carbon dioxide is fixed and 6 carbon sugar is formed.

Calyculus : The collective term for the involucral bracts (or phyllaries) surrounding a head in the Asteraceae.

Calyptra (Operculum) : A cap-like covering or lid of some flowers or fruits that becomes detached at maturity by abscission; e.g. (1) the cap on the buds of eucalypts, (2) the lid of circumsciss capsules.

Calyptra : Small sheath of cells found in non-vascular plants, derived from the archegonium to cover the tip of the capsule partially or completely.

Calyx : The outer series of floral leaves forming the perianth of a flower; made up of the sepals, which are green and encircle the flower bud.

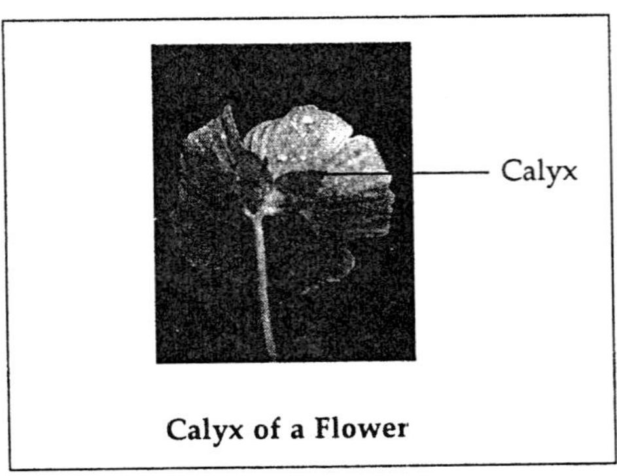

Calyx of a Flower

Calyx Limb : The structure formed by the limbs of all the sepals.

Calyx Tube : A tube formed by the fusion of the sepals, but sometimes wrongly used in the sense of hypanthium.

CAM Plant : Standing for crassulacean acid metabolism, this is a type of plant that employs an alternative photosynthesis pathway where CO_2 enters the open stomata of the leaf during the night, allowing the stomata to close during the day to reduce water loss. This adaptation is useful in very hot, arid climates. The pineapple is an example of this type of plant.

Cambium : Layer of meristematic tissue (also known as lateral meristems), responsible for secondary growth.

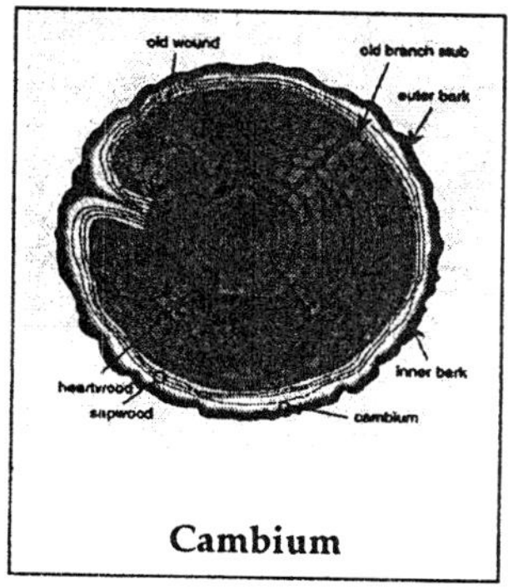

Cambium

Camouflage : A feature common to invertebrates, which helps them blend with their surroundings using its skin colours or patterns.

Canaliculate : Longitudinally channelled or grooved.

Cancer : Diseases in which abnormal cells divide and grow unchecked. Cancer can spread from its original site to other parts of the body and can be fatal.

Candidate Gene : A gene located in a chromosome region suspected of being involved in a disease.

Canescent : With gray or white short hairs, often having a hoary appearance.

Canopy : (1) the branches and foliage of a tree; (2) often used as a collective term for the crowns of trees in a forest.

Capillary : Very slender and hairlike.

Capillary Array : Gel-filled silica capillaries used to separate fragments for DNA sequencing. The small diameter of the capillaries permit the application of higher electric fields, providing high speed, high throughput separations that are significantly faster than traditional slab gels.

Capillary Water : Water held in the tiny pores between soil particles by the adhesive force : surface tension.

Capitate : (1) shaped like a head; (2) in a head-like cluster.

Capitiform : Shaped like a head, somewhat globose.

Capitulescence : A special term used in Asteraceae to describe a group of associated heads—also called capitula; it is analogous to an inflorescence)

Capitulum (head) : A dense cluster of more or less sessile flowers, e.g. in Asteraceae a group of florets sessile on a common receptacle.

Capsid : The outer proteinaceous coat of a virus.

Capsomere : A protein sub-unit of the capsid of a virus.

Capsule : A dry, generally many-seeded fruit divided into two or more seed compartments that dehisces or splits open longitudinally with the line of dehiscence either through the locule (loculicidal) or through the septa (septicidal), or, less commonly, through pores (poricidal) or around the circumference (circumscissile)

Carbon Cycle : The cycle where carbon-dioxide is taken in and converted to organic compounds by photosynthesis or chemosynthesis, after which it is partially incorporated into sediments, and then returned to the atmosphere by respiration or combustion.

Carbon Fixation : Conversion of carbon-dioxide and other single carbon compounds to organic compounds such as carbohydrates.

Carbon-Nitrogen (C/N) Ratio : Ratio of carbon mass to nitrogen mass in soil or other organic material.

Carboxyl Group : The COOH group found attached to the main carbon skeleton in certain compounds, like carboxylic acids and fatty acids.

Carboxysomes : Polyhedral cell inclusions which form the key enzyme of the Calvin cycle.

Carcinogen : An often mutated substance which is implicated as one of the causing agents of cancer.

Carina : A keel; hence carinate, longitudinally keeled.

Carinate : Keeled with one or more longitudinal ridges.

Carnasial Tooth : A premolar tooth, which is used to efficiently tear and slice meat of prey. This tooth is seen only in the carnivores.

Carotenoids : Yellow and orange pigments that complement the green pigments of chlorophyll by absorbing wavelengths of light that chloropyhll can not.

Carpel : A unit of the female part of the flower (gynoecium), consisting of an ovary bearing one or more ovules, a receptive stigma, and often a stalklike style between them. A flower can have a solitary carpel or more than one carpel. If the carpels (pistils) are free the gynoecium is apocarpous or if the carpels are fused the gynoecium (pistil) is syncarpous (or compound).

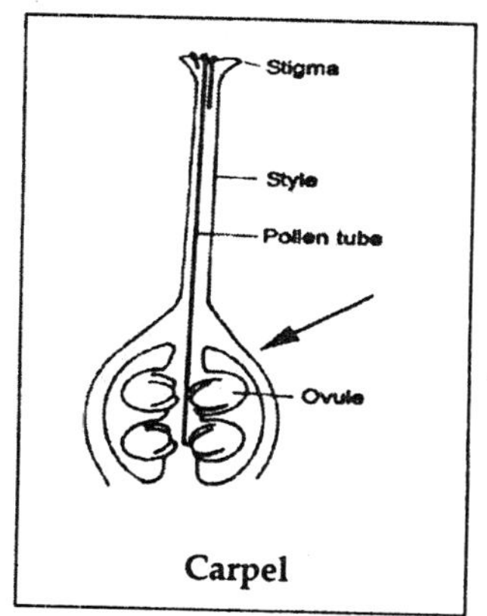

Carpel

Carpogonium : Haploid (N) female sexual structure in Rhodophyta consisting of single cell and its extension, the trichogyne.

Carpophore : In a fruit, the stalk of a mericarp.

Carposporangia : Single celled structures that produce the diploid (2N) carpospores on the parasitic carposporophyte in Rhodophyta.

Carpospore : Non-motile diploid (2N) spores formed on parasitic carposporophyte stage of the Rhodophyta, which germinate to form tetrasporophyte stage.

Carposporophyte : The diploid (2N) stage of Rhodophyta which develops after fertilisation of the carpogonium.

Carrier : An individual who possesses an unexpressed, recessive trait.

Carrying Capacity : It is the maximum population of a particular species, which can be supported for an indefinite period of time in a particular environment.

Cartilaginous : Hard, tough, without chlorophyll and vasculature; like cartilage.

Caruncle (strophile) : An appendage of a seed, near the hilum (scar). adj. caruncluate.

Caryopsis : Small, dry, single seeded fruits which do not split at maturity. The pericarp cleaves to the seed coat; typically seen in grains.

Casparian Strip : Band of cell wall material in the radial and transverse walls of the endodermis. It stops the passive flow of materials into the stele.

Casque : A formation on the head resembling a helmet, that is located on the head of a lizard.

Caste : A group of species, which shares similar features, form or behaviour and belong to the same social group.

Catabolic Pathway : A type of metabolic pathway whereby large, complex molecules are broken down into smaller ones in order to release energy.

Catabolism : A process by which complex substances are broken down into simpler compounds, often accompanied by the release of energy.

Catabolite Repression : Transcription-level inhibition of inducible enzymes by glucose, or other easily available carbon sources.

Catadromic, Catadromous : With the first sub-branch of a lateral branch produced on the basiscopic margin, mostly of venation in pinnate ferns.

Cataphyll : (1) a scale leaf associated with a vegetative part of a plant; (2) a leaf composed mostly of a leaf sheath or base with the lamina reduced to a minute awn, e.g. in some Juncus species.

Cation Exchange : Replacement of an essential element cation released from a soil particle by a proton.

Catkin : A spikelike, often pendulous, inflorescence of petalless unisexual flowers, either staminate or pistillate.

Caudate : Tailed excessively acuminate so that the tip is long and weak.

Caudex : The persistent, often woody base of an otherwise annual herbaceous stem.

Cauliflory : The production of flowers or fruits on well-developed trunks or major branches. adj. cauliflorous.

Cauline : Attached to or referring to the stem, as opposed to 'basal', often used to describe leaf position.

Cavitation : The rupture of the water column in the xylem, when tension surmounts the cohesive nature of water.

CDNA Library : A collection of DNA sequences that code for genes. The sequences are generated in the laboratory from mRNA sequences.

Cell : Microscopic structure forming the basic structural and functional unit of living organisms. It encompasses nuclear and cytoplasmic material enclosed by a cell membrane.

Cell Biology : Branch of biology involving the study of cells, their structure, formation, components and functions.

Cell Cycle : The complete sequence of events of a dividing cell.

Cell Division : Process of division of cell with the purpose of growth or reproduction.

Cell Membrane : The semipermeable membrane sheathing cytoplasmic material of the cell.

Cell Plate : During cell division, the plate formed at the midpoint between two sets of chromosomes, which is involved in the wall formation between two daughter cells.

Cell Sap : Fluid present in the central vacuole of plant cells.

Cell Wall : A protective layer formed outside of the plasma membrane in plant cells.

Cell-mediated Immunity : Immunity resulting from destruction of foreign organisms and infected cells by the active action of T-lymphocytes on them. It can be acquired by individuals by the transfer of cells.

Cellular Slime Molds : Slime molds with a vegetative phase containing amoeboid cells that come together to form a pseudoplasmodium.

Cellulitis : A diffused inflammation of the soft or connective tissue, in which a thin and watery exudate spreads through tissue spaces, often leading to ulceration and abscess formation.

Cellulose : A complex carbohydrate composed of glucose units, which forms the major constituent of cell wall in plant cells.

Celsius : A temperature measurement scale. Water freezes at 0°C, water boils at 100°C. °C= (5/9)(°F-32).

Centifugal : Tending outwards or developing from the centre outwards.

Centimorgan : A unit of measure of recombination frequency. One centimorgan is equal to a 1% chance that a marker at one genetic locus will be separated from a marker at a second locus due to crossing over in a single generation. In human beings, one centimorgan is equivalent, on average, to one million base pairs.

Central Cell Nuclei : Mostly two in number nuclei uniting with sperm to form primary endosperm nucleus in embryo sac. It is a membrane enclosed organelle of eukaryotic cells that contains its genetic material in the form of chromosomes.

Centri : Prefix pertaining to the centre of an organ.

Centrifugal : Developing from the centre outward.

Centrioles : Small, cylindrical cell organelles found in animals and some algae and fungi. Located near the nucleus in the cytoplasm of most eukaryotic cells each centriole is usually composed of nine triplets of microtubule.

Centripetal : Tending inwards or developing towards the centre from the outside.

Centromere : Portion of the chromosome holding the two chromatid together before anaphase stage of mitosis or anaphase II stage of meiosis. The spindle fibres are attached to this region and move the chromosomes during cell division.

Cephalosporin : A group of broad-spectrum, penicillinase resistant antibiotics, derived from *Cephalosporium*.

Cespitose : Having a densely clumped, tufted or cushion-like growth form with the flowers extending above the clump.

Chaff : (1) membranous scales or bracts; (2) thin dry unfertilised ovules among the fully developed seeds of a fruit, as in many eucalypts.

Chambered : Said of pith which is interrupted by hollow spaces.

Chaparral : A type of biome characterised by evergreen shrubs, rainy winters, long dry summers, and cycles of fire burnoff.

Chaperonin : Heat shock proteins that oversee correct folding and assembly of polypeptides in bacteria, plasmids, eukaryotic cytosol and mitochondria.

Character : Any feature of an organism or taxonomic group that can be measured, counted or otherwise assessed.

Character Displacement : Adaptations of different sets of characteristics in two similar species, brought about by overlapping territories, resulting in competition.

Chasmogamous : Of flowers that are pollinated while open.

Chemiosmosis : A term used to describe the ability of a membrane to use energy to pump hydrogen ions and then harness the hydrogen ion gradient to carry out cellular functions, most importantly, the production of ATP's.

Chemiosmotic Phosphorylation : Occurring in mitochondria and chloroplasts, this process involves the synthesis of ATP from ADP and phosphate unit.

Chemoautotroph : A type of organism that needs CO_2 for their carbon source, but obtains energy from inorganic molecules.

Chemoheterotroph : A type of organism that must consume organic molecules for their carbon source and to obtain energy.

Chemolithotroph : Living organisms that obtain their energy from oxidation of inorganic compounds, which act as electron donors.

Chemoorganotroph : Organisms that obtain energy and electrons from the oxidation of organic compounds.

Chemostat : A continuously used culture device, controlled by limited amounts of nutrients and dilution rates.

Chemosynthetic Origin of Life : Theory according to which life began via a series of chemical reactions on primitive Earth.

Chemotaxis : Movement of a motile organism under the influence of a chemical. It maybe attracted towards the chemical or maybe repulsed by it.

Chemotrophs : Organisms that obtain their energy by the oxidation of chemical compounds.

Chiasma : X-shaped structure formed by the attachment of two chromatids of homologous chromosomes to each other during meiosis.

Chimera (pl. Chimaera) : An organism that contains cells or tissues with a different genotype. These can be mutated cells of the host organism or cells from a different organism or species.

Chimeraplasty : An experimental targeted repair process in which a desirable sequence of DNA is combined with RNA to form a chimeraplast. These molecules bind selectively to the target DNA. Once bound, the chimeraplast activates a naturally occurring gene-correcting mechanism. Does not use viral or other conventional gene-delivery vectors.

Chitin : Polymer composed of partly amino sugars, it is a semitransparent hard substance forming the outer covering or exoskeleton of crustaceans, arachnids and insects.

Chlamydospore : A thick walled intercalary or terminal asexual spore which is not shed. It is formed by rounding up of a cell.

Chlorenchyma : Parenchyma tissues with chlorophyll content.

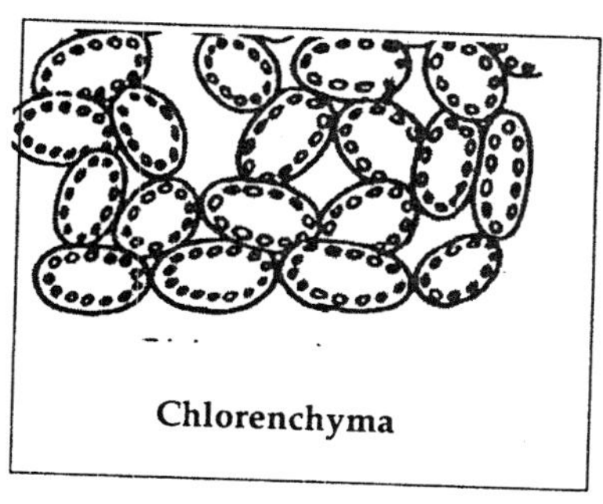
Chlorenchyma

Chlorophyll : Green pigment found in plants, cyanobacteria and algae, which is involved in capturing light energy required for photosynthesis.

Chloroplast : The organelle in plants and some protests that absorbs sunlight and uses the energy to synthesis organic compounds from CO_2 and H_2O.

Chloroplast Chromosome : Circular DNA found in the photosynthesising organelle (chloroplast) of plants instead of the cell nucleus where most genetic material is located.

Chlorosis : A plant disorder commonly caused by a deficiency in iron which results in the yellowing of leaves (except for the veins). High alkaline soils can hinder a plant's ability to uptake iron, so there may not be a problem with the iron content of the soil itself.

Chromatid : One of the two identical chromosome strands united by a centromere into which the chromosome longitudinally splits while preparing for cell division.

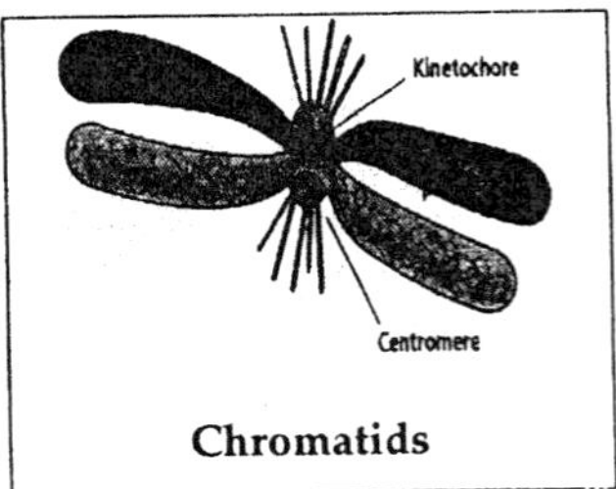

Chromatids

Chromatin : Found in chromosomes, chromatin is a readily staining substance of a cell nucleus containing DNA, RNA and other proteins that form chromosomes during cell division.

Chromista : A sometimes recognised Kingdom that includes brown algae, golden algae, and diatoms.

Chromomere : One of the serially aligned beads or granules of a eukaryotic chromosome, resulting from local coiling of a continuous DNA thread.

Chromoplast : Plastids containing pigments other than chlorophyll, usually imparting red or yellow colour.

Chromosomal Deletion : The loss of part of a chromosome's DNA.

Chromosomal Inversion : Chromosome segments that have been turned 180 degrees. The gene sequence for the segment is reversed with respect to the rest of the chromosome.

Chromosome : The self-replicating genetic structure of cells containing the cellular DNA that bears in its nucleotide sequence the linear array of genes. In prokaryotes, chromosomal DNA is circular, and the entire genome is carried on one chromosome. Eukaryotic genomes consist of a number of chromosomes whose DNA is associated with different kinds of proteins.

Chromosome Condensation : Also called pachytene, this process is a part of prophase I, wherein the chromosomes become shorter and thicker.

Chromosome Painting : Attachment of certain fluorescent dyes to targeted parts of the chromosome. Used as a diagnositic for particular diseases, e.g. types of leukemia.

Chromosome Region p : A designation for the short arm of a chromosome.

Chromosome Region q : A designation for the long arm of a chromosome.

Chronic Carrier : An individual carrying a pathogen over an extended period of time.

Chytrid : A fungus belonging to the genus Chytridomycota. It is spherical in shape and has rhizoids, which are short, thin filamentous branches, that resemble fine roots.

Cilia : Minute hairlike extensions present on a cell surface, which move in a rhythmic manner.

Ciliate : With a row of fine hairs situated along the margin of a structure such as a leaf

Ciliolate : Having the margin minutely fringed.

Cilium : Precisely arranged, short microtubules found mostly in bunches, similar to a flagellum. These may either be sensory or locomotory organelles.

Ciliun : A hair or bristle; hence ciliate, usually with marginal hairs.

Cinereous : Ash-coloured, light-gray due to a covering of short hairs.

Circadian Rhythm : An rhythmic daily activity cycle exhibited by many organisms in an intervals of 24 hours.

Circinate : Spirally coiled with the tip innermost as in the young fronds of many ferns.

Circinnate : Coiled, with the apex innermost (n. circinnus), as in many developing fern fronds.

Circular (Orbiculate) : A 2-dimensional shape with length and breadth more or less equal

Circumboreal : Distributed around the globe at northern latitudes.

Circumsciss (Circumscissile) : Breaking open along a transverse line so that the top (calyptra) comes off like a lid, as in some capsules.

Circumsessile : Dehiscing along a transverse circular line around the fruit or anther, so that the top separates or falls off like a lid.

Cirrhose **:** Tendril-like, with a slender, coiled or wavy tip.

Cismontane **:** Referring to the ocean-facing side as opposed to the desert-facing side of the mountains.

Citric Acid Cycle **:** In aerobic respiration, the complex series of reactions following glycolysis, which involve mitochondria, ATP and enzymes.

Cladistics **:** It is the study of evolutionary history of a group of organisms, especially as shown in a family tree.

Cladode **:** A photosynthetic stem whose foliage leaves are usually reduced or absent.

Cladophyll **:** Also called phylloclade, this is a flattened stem that looks like a leaf.

Clarification **:** The process of purification of water, where suspended material in the water is removed. It can be done by using sedimentation, filtration or by the use of adsorbing chemicals like alum.

Clasping **:** Surrounding or embracing, e.g. stem-clasping, of a lamina surrounding the stem.

Class **:** In classification, the category coming between a division and order.

Clathrate **:** In the form of a lattice; used to describe scales with a single layer of translucent cells with dark cross-like walls.

Clavate **:** Club-shaped, gradually thickened or widened toward the apex.

Clay Soil **:** The heaviest soil classification, composed of closely packed particles that allow less water and air movement through the soil. Clay soils cool slowly, warm slowly, and can hold large volumes of nutrients.

Cleft **:** Deeply cut, usually more than one-half the distance from the margin to the midrib or base.

Cleistogamous **:** Of flowers that remain closed and are self-pollinating and set fertile seed.

Climograph **:** Annual cycle of temperature and rainfall for a particular geographical area depicted in a graphical format.

Clone : Cells which have descended from a single parent cell. Organisms having identical copies of DNA structure, which is obtained by replication.

Cloning : Using specialised DNA technology to produce multiple, exact copies of a single gene or other segment of DNA to obtain enough material for further study. This process, used by researchers in the Human Genome Project, is referred to as cloning DNA. The resulting cloned (copied) collections of DNA molecules are called clone libraries. A second type of cloning exploits the natural process of cell division to make many copies of an entire cell. The genetic make-up of these cloned cells, called a cell line, is identical to the original cell.

Cloning Vector : DNA molecule originating from a virus, a plasmid, or the cell of a higher organism into which another DNA fragment of appropriate size can be integrated without loss of the vector's capacity for self-replication; vectors introduce foreign DNA into host cells, where the DNA can be reproduced in large quantities. Examples are plasmids, cosmids, and yeast artificial chromosomes; vectors are often recombinant molecules containing DNA sequences from several sources.

Close : In reference to bark means the bark is tight, not peeling or shedding, not scaly.

Closed Carpel : Is another phrase used for Angiosperms that are plants with seeds inside the ovary.

Closed Forest (Rainforest) : A forest dominated by broad-leaved trees with dense crowns that form a continuous layer (canopy) and with one or more of the following growth forms.

Clutch : Eggs or young of a species produced in single a breeding attempt by a female.

Cobwebbed : Covered with long weak, loosely entangled hairs, resembling a spiderweb; usually whitish. adj. cobwebby.

Coccus : One of the segments (usually 1-seeded) of a distinctly lobed fruit which becomes separated at maturity; sometimes called a mericarp.

Codominance **:** Situation in which two different alleles for a genetic trait are both expressed.

Codon **:** Triplet of adjacent nucleotides in messenger RNA, which specify the amino acid to be incorporated into a protein.

Coenocytic **:** Large cells containing myriad nuclei. It is formed when the cell nucleus divides multiple times without the actual division of the cell.

Coenosorus **:** An extended sorus or series of sori that have united to coalesced.

Co-enzymes **:** Molecules providing transfer site for biochemical reactions catalysed by enzymes.

Cohesion **:** The sticking together of two or more similar parts that are not organically fused. adj. coherent.

Cohesion, Hence Coherent **:** Of two similar organs or parts touching one another, +/- adhesively but easily separated and not fused or grown together.

Cohesion-Tension Theory **:** This theory explains that the upward pull of water takes place by the combination of water molecules cohesion in the vessels and tracheids and tension on the water column caused by transpiration.

Coisogenic or Congenic **:** Nearly identical strains of an organism; they vary at only a single locus.

Coleoptile **:** The first leaf above ground level forming a sheath around the tip of the stem, so as to protect the emerging shoot (plumule) of monocotyledons like grasses and oats.

Coleorhiza **:** Sheath formed around the emerging radicle in plants of the monocotyledons like the grass family.

Collar **:** In grasses the outer side of the leaf at the junction of the sheath and blade.

Collateral Bud **:** An accessory bud adjacent to or at the side of the main axillary bud.

Collenchyma **:** Cells containing primary walls thickened at the cells corners, but thin elsewhere.

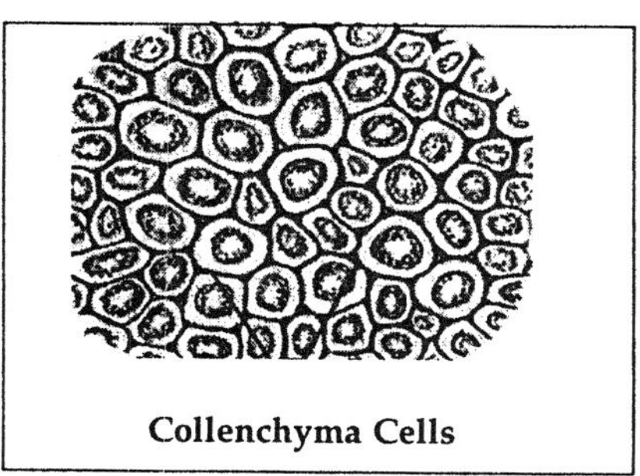

Collenchyma Cells

Colliculate **:** Of a surface, rough with low rounded protuberances.

Colonial **:** Forming colonies; used mainly for plants with underground connections.

Colonisation **:** Establishment of an entire community of microorganisms at a designated site.

Colony **:** Aggregation or loose organisation of similar cells (or organisms) which developed from a single original cell or organism, each capable of independence from the rest of the colony.

Colourless Sulfur Bacteria **:** A group of nonphotosynthetic bacteria that oxidise sulfur compounds, thus deriving their energy by this process.

Colt **:** Male horse less than four years of age.

Columella **:** A small column of tissue which runs up through the centre of a spore capsule. It is present in hornworts, mosses, and some rhyniophytes.

Column **:** (1) (gynostemium) a structure formed by the union of stamens, style and stigmas, as in Orchidaceae, Asclepiadaceae and Stylidiaceae; (2) the lower part of an awn in grasses, when different in form from the upper part.

Coma **:** A tuft of hairs, especially on a seed or fruit.

Combination **:** Taxonomically the name of a genus combined with a specific epithet.

Combinatorial Biology : The process of transfer of genetic material from one microorganism to another. Mostly used to synthesise products such as antibiotics. It is also used in genetic engineering.

Cometabolism : Transformation of a substrate by a microorganism without deriving energy or nutrients from the substrate.

Commensal : Organisms living in, on or with another, without any particular advantage or harm from the relationship.

Commensalism : A relationship between organisms whereby one of them benefits while the other is neither helped nor hurt.

Community : An assemblage of plants living in the same place.

Companion Cells : Specialised parenchymal cells situated beside sieve tubes in the phloem of angiosperms that regulate flow of nutrients through the sieve tube.

Comparative Genomics : The study of human genetics by comparisons with model organisms such as mice, the fruit fly, and the bacterium *E. coli.*

Compensation Point : Conditions where respiration is in balance with photosynthesis, with generally no net gas exchange, in algae usually related to habitat depth.

Competent : The ability to take up DNA.

Competition : When two or more individuals compete for the same set of available and limited resources, affecting both the parties negatively.

Complementary DNA (cDNA) : DNA that is synthesised in the laboratory from a messenger RNA template.

Complementary DNA : A DNA copy of any RNA molecule, like mRNA or tRNA.

Complementary Sequence : Nucleic acid base sequence that can form a double-stranded structure with another DNA fragment by following base-pairing rules (A pairs with T and C with G). The complementary sequence to GTAC for example, is CATG.

Complete : Describing flowers that contain petals, sepals, pistils and stamens.

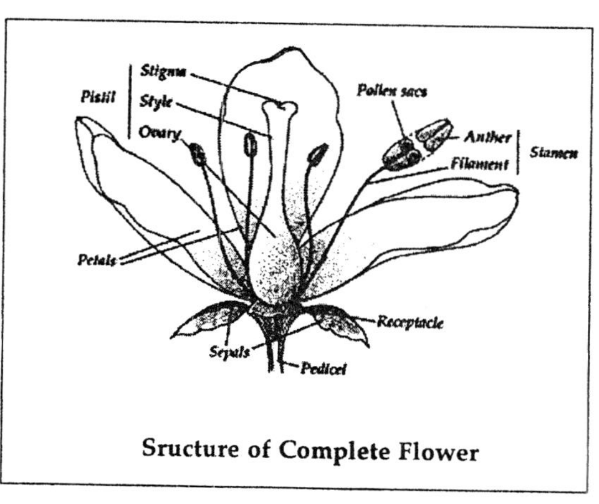

Sructure of Complete Flower

Complex Trait : Trait that has a genetic component that does not follow strict Mendelian inheritance. May involve the interaction of two or more genes or gene-environment interactions.

Complex Viruses : Viruses with capsids that are neither icosahedral nor helical. They have a complicated symmetry.

Composite : A member of the Asteraceae, or sunflower family, previously called the *Compositae*.

Compost : Combination of several dead and decaying organic substances, such as manure, dead leaves, etc. used for soil fertilisation.

Compound : Composed of several more or less similar parts, as opposed to simple; e.g. of an ovary formed from severalunited carpels or of a leaf divided into leaflets.

Compound Corymb : One in which some of the pedicels branch again in the same way that the pedicels of a simple corymb branch.

Compound Leaf : One divided into separate leaflets, i.e., one in which the blade consists of two or more separate leaflets on a common leaf-stalk or rachis.

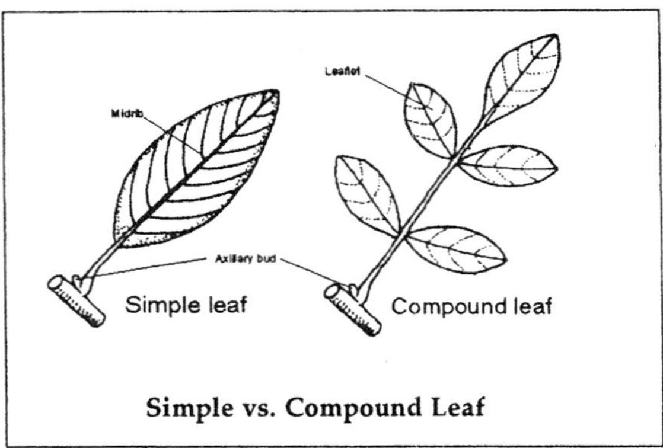

Simple vs. Compound Leaf

Concave : Saucer-shaped, with surface that curves inward.

Conceptacle : Cavity or chamber containing reproductive structures with opening to the surface for release, e.g. in Corallinales and Fucales.

Concolorous : With the same colour throughout or on both surfaces.

Concrescent : Growing together; hence concrescence.

Conditional Mutations : Mutations occurring only under certain specific conditions.

Conditioning : A learning method either using a stimulus–response, or a reward-punishment method, in which associations are made.

Conduplicate : Folded lengthwise with the upper surface inwards, eg on a leaf folded along the midrib.

Cone : Geometrically a solid, circular in cross-section, triangular in longitudinal section, hence conical; botanically a series of spirally arranged wooden scales or sporophlls.

Confidentiality : In genetics, the expectation that genetic material and the information gained from testing that material will not be available without the donor's consent.

Conflorescence : A branch system bearing flowers in which the main axis bears uniflorescences, but is itself qualitatively different in structure from the uniflorescences.

Confluent : Running together or blending of one part into another.

Conform : Of the same or similar shape, eg pinnate fronds with the apical pinnae similar to the lateral.

Congeneric : Belonging to one and the same genus; hence congener (n.).

Congenital : Any trait present at birth, whether the result of a genetic or nongenetic factor.

Conical : Of a solid in the form of a cone, attached be the broad end.

Conidiospore : A thin-walled, asexual spore seen on hyphae which is not contained in sporangium.

Conidium : Fungal spore formed outside a sporangium and produced asexually.

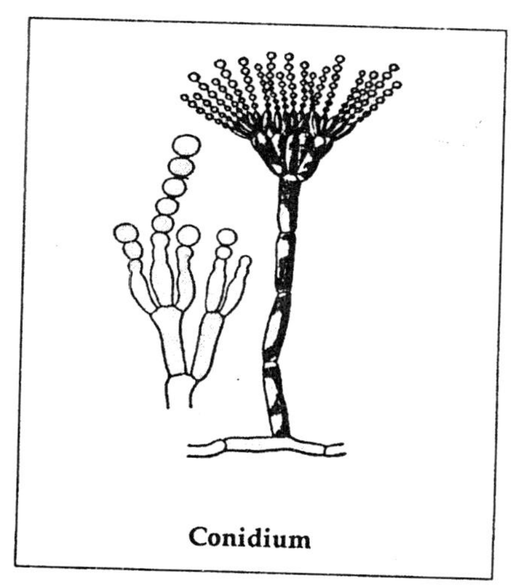

Conidium

Conifer : A plant commonly having needlelike, persistent leaves and a woody cone for a fruit.

Conjugants : Mating partners that participate in conjugation, which is a type of sexual reproduction, seen in protozoans.

Conjugation : Process of genetic exchange occurring in bacteria and some green algae, wherein the DNA is passed through a tube connecting adjacent cells.

Conjugative Plasmid : A self transmissible plasmid, or a plasmid that can encode all functions required to bring about its conjugation.

Connate : Describing similar structures that are joined or grown together.

Connective : The sterile part of an anther connecting the loculi.

Connivent : Converging together, usually of organs with their bases separate and their appices approaching each other, not touching or fused.

Conserved Sequence : A base sequence in a DNA molecule (or an amino acid sequence in a protein) that has remained essentially unchanged throughout evolution.

Consortium : Two or more members working together, where each organism benefits from the other, thus often performing functions that may not be possible to carry out individually.

Conspecific : Belonging to one and the same species.

Constitutive Ablation : Gene expression that results in cell death.

Constitutive Enzyme : Enzymes synthesised in the cell, irrespective of the environmental conditions surrounding the cell.

Constriction : This is a method used by nonvenomous snakes to tightly grip and suffocate their prey, by coiling around the prey.

Consumer : An organism, often an animal, which feeds on plants or other animals.

Contig : Group of cloned (copied) pieces of DNA representing overlapping regions of a particular chromosome.

Contig Map : A map depicting the relative order of a linked library of overlapping clones representing a complete chromosomal segment.

Contigous : Adjacent and touching but not united.

Contorted : Twisted; a form of imbricate aestivation in which each segment has one edge overlapping the next segment.

Contracted : Narrowed or shortened as opposed to open or spreading.

Convergent : Growing or lying towards one another.

Convergent Evolution : Similar structural appearance in organisms, which have different lines of descent.

Convex : Having surface that curves outward.

Convolute : Rolled up longitudinally, with one edge inside the other and the upper surface on the inside.

Copious : Very much, very many.

Coppice Shoot : A shoot developed from a dormant bud in the trunk or larger branches of a tree, the leaves on such a shoot often differ from the adult leaves and are called juvenile leaves (similar to sapling leaves); a common feature of many eucalypts and rainforest trees. Coppiceshoots usually develop after damage to the trunk by fire, cutting etc.

Cork : Outer tissue layer of an old woody stem produced by cork cambium, whose cells are saturated with suberin at maturity.

Cork Cambium : Lateral meristematic tissue ring found in woody seed plants between the exterior of woody stems or roots and central vascular tissue. It produces cork to its exterior and phellogen to its interior.

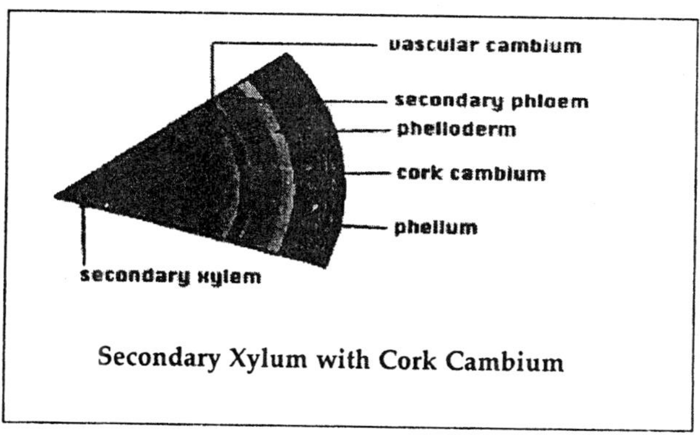

Secondary Xylum with Cork Cambium

Corm : A short, broad, fleshy, subterranean stem which produces aerial stems, leaves and flowers and in which food reserves are stored.

Corniculate : Having little horns or hornlike appendages.

Corolla : The inner whorl of the perianth, between the calyx and the stamens, a collective term for the petals of a flower.

Corona : Any ring of tissue or appendage that stands between the perianth and the stamens, or on the perianth. e.g. as in Passiflora species.

Cortex : Generally parenchyma cells forming a tissue extending between the vascular tissue and epidermis.

Corymb : A broad, flat-topped inflorescence in which the flower stalks arise from different points on the main stem and the marginal flowers are the first to open.

Corymb Iinflorescence

Cosmid : Artificially constructed cloning vector containing the cos gene of phage lambda. Cosmids can be packaged in lambda phage particles for infection into *E. coli*; this permits cloning of larger DNA fragments (up to 45kb) than can be introduced into bacterial hosts in plasmid vectors.

Costa : Rib, especially the midrib of a leaf or pinna; hence costate (pl. costae; dimin. costule).

Costal : Pertaining to or near the costa (dimin. costular).

Costate : Ribbed, having longitudinal elevations.

Costule : Midrib of higher order pinnule or lobe; hence costulate.

Cotyledon : n. The "seed leaves" produced by the embryo of a seed plant that serve to absorb nutrients packaged in the seed, until the seedling is able to produce its first true leaves and begin photosynthesis; the number of cotyledons is a key feature for the identification of the two major groups of flowering plants.

Countershading : The development of dark colours on the areas exposed to the sun and light colours on the undercarriage.

Cover Cells : Sterile cells cut off from during the formation of tetrasporangia in Rhodophyta.

Crater : A bowl; hence craterous, crateriform.

Crenate : With shallow roundish or bluntish teeth on the margin, scalloped.

Crested : With an elevated ridge or line along the summit of an organ.

Crisped : With the margins finely wavy, curled or crumpled.

Cristate : With an appendage resembling a crest.

Critical Habitat : A habitat which is critical for the survival and conservation of a species, designated by a rule published in the Federal Register.

Crossing Over : The breaking during meiosis of one maternal and one paternal chromosome, the exchange of corresponding sections of DNA, and the rejoining of the chromosomes. This process can result in an exchange of alleles between chromosomes.

Crosswall : End wall of a cell in a filament.

Crown : Part of a tree, shrub, etc, above the first branching.

Crown Division : Asexual type of reproduction, involving the division of the base of the stem.

Crozier : The coiled young fronds of ferns.

Cruciate : Generally cross-shaped; in the Rhodophyta, the type of division where tetrasporangium divides in two planes at right angles to each other.

Cruciate, Cruciform : In the shape of a cross.

Cryptogam : Plants reproducing by spores and not seeds, eg. ferns, mosses, fungi etc.

Cryptostoma : Cavity open to surface that contains hairs or pariphyses, e.g. in Fucales.

Cucullate : Hooded or hood-shaped.

Culm : An aerial stem bearing the inflorescence, in grasses, rushes etc.

Cultigen : Plant known only in cultivation, apparently originating under domestication.

Cultivar : Cultivated variety, a variety developed incultivation. An assemblage of cultivated individuals distinguished by any characters significant for the purposes of agriculture, forestry or horticulture, and which, when reproduced retains its distinguishing features.

Cuneate : Wedge-shaped, with the narrow part at the point of attachment.

Cup or Cupule : That in which an acorn rests.

Cupule : A cuplike structure the subtends flower or fruit.

Cusp : A sharp, rigid point; hence cuspidate, with the apex abruptly narrowed into a point.

Cuspidate : Tipped with an abrupt short, sharp, firm point.

Cuticle : Thin hyaline film derived from the exterior surfaces of epidermal cells, covering the surface of plants.

Cutin : Fatty or waxy substance making up the cuticle.

Cutting : Any part of a plant that is cut off from the parent plant and rooted to form a new plant. Types include : root cutting, leaf cutting, and stem cutting.

Cyanobacterium : A photosynthetic, nitrogen fixing bacteria which includes the blue-green bacteria.

Cyathium : The specialised inflorescence characteristic of the *Euphorbiaceae*, consisting of a flower-like, cup-shaped involucre which carries the several true flowers within.

Cyclosis : Flow of cytoplasm with the cell.

Cylindric : Tubular or rod-shaped.

Cyme : An inflorescence in which the main axis ends in a flower and further growth of the inflorescence is by one or more branches which themselves end in a flower (the main and lateral branches may bear bracteoles but have no bracts, leaves or nodes). adj. cymose. e.g. dichasium, monochasium.

Cypsela : The fruit formed in most Asteraceae; a dry indehiscent 1-seeded fruit formed from inferior ovary of 1 carpel, with the seed not fused to the fruit wall and usually topped by the pappus.

Cyst : Resting stage of certain bacteria and protozoans, wherein the entire cell is surrounded by a protective layer.

Cystocarp : The carposporophyte and surrounding envelop of haploid (1N) female gametophyte tissue (or pericarp) in the Rhodophyta.

Cystocarpic : Bearing cystocarps, signaling a fertilised female gametophyte in the Rhodophyta.

Cystoliths : Mineral concretions, usually of calcium carbonate on a cellulose stalk, chiefly occurring in specialised hairs in some Urticaceae and Cannabaceae and in Acanthaceae; often appearing as tubercle-based hairs.

Cytochrome : Protein containing iron, acting as small electron carriers by transferring molecules in electron transport system.

Cytogenetics : Study of genetic effects of chromosome behaviour and structure.

Cytokine : Non-antibody proteins released by a cell when it comes in contact with specific antigens.

Cytokinesis : Process of cell division, as opposed to nuclear division.

Cytokinin : Growth hormone concerned with cell division and other metabolic activities of the cell.

Cytological Band : An area of the chromosome that stains differently from areas around it.

Cytological Map : A type of chromosome map whereby genes are located on the basis of cytological findings obtained with the aid of chromosome mutations.

Cytoplasm : The protoplasm of a cell, exclusive of the nucleus.

Cytoplasmic Membrane : A selectively permeable membrane which is present around the cytoplasm of the cell.

Cytoplasmic Trait : A genetic characteristic in which the genes are found outside the nucleus, in chloroplasts or mitochondria. Results in offspring inheriting genetic material from only one parent.

Cytosine *(C)* : A nitrogenous base, one member of the base pair GC (guanine and cytosine) in DNA.

Cytoskeleton : Network of microfilaments and microtubules that protects and maintains cell shape, enables cellular motion and plays important roles in both intracellular transport and cell division. It is present in both prokaryotes and eukaryotes.

Cytosol : Fluid part of the cell into which the organelles are scattered.

D

Dark Reactions : Stage of photosynthesis which is light independent wherein carbohydrates are synthesised from carbon dioxide.

Data Warehouse : A collection of databases, data tables, and mechanisms to access the data on a single subject.

Day-neutral Plant : Plants independent of specific day lengths for commencement of flowering.

Deadhead : Take off dead flowers. This is done to prolong plant life in annuals by preventing seed formation via flowering.

Death Rate : The average number of newborn or young ones dying within a specified period of time.

Deciduous : (1) falling seasonally, e.g. of the leaves and bark of some trees, caducous; (2) a plant losing its leaves for part of the year.

Decomposer : Organism breaking down organic matter into forms suitable for recycling.

Decomposition : Chemical breakdown of a compound into smaller and simpler compounds by microorganisms.

Decompound : Several times divided or compound.

Decumbent : Spreading horizontally with the ends growing upwards.

Decurrent : Extending downward from the point of insertion; said of a leaf which extends down the stem (or twig) below the point of fastening or where the blade extends down the petiole.

Decurved : Bent downwards and curved.

Decussate : Arranged in pairs along the stem with each pair at right angles to the one above and below.

Dedifferentiate : Pertaining to cells, dedifferentiate means becoming less specialised.

Defined Medium : A medium whose quantitative and chemical composition is exactly known.

Definite : Of precise and constant number or extent.

Deflexed : Bent downwards towards the base of the stem.

Defoliation : The loss of foliage due to unnatural causes, such as chemicals, frost, or high winds.

Degradation : Process by which a compound is transformed into simpler compounds.

Dehisce : To split open; hence dehiscence, dehiscent.

Dehiscent : Opening spontaneously when ripe to discharge the seed content.

Deletion : A loss of part of the DNA from a chromosome; can lead to a disease or abnormality.

Deletion Map : A description of a specific chromosome that uses defined mutations —specific deleted areas in the genome— as 'biochemical signposts,' or markers for specific areas.

Deliquescent : Branching in such a way that the stem is lost in the branches; the main stem branching into numerous smaller ones, e.g., American elm is a deliquescent tree branching without a continuous main stem.

Delphinidae : A group of marine mammals that belong to the family Delphinidae and the Order Cetacea, like dolphins and their relatives.

Deltate : A 2-dimensional triangular shape.

Deltoid : Of a triangular shape, like the Greek capital letter delta (A)

Deme : A local breeding that interbreeds organisms of the same species or individuals.

Denaturation : Process by which double stranded DNA unwinds into two single strands.

Dendriform, Dendroid : Treelike, mainly of branching and form but not in size.

Dendritic : Much branched, like the crown of a tree, e.g. of hairs.

Denitrification : Reduction of nitrate or nitrite into simpler nitrogenous compounds like molecular nitrogen or nitrogen oxides.

Dense : Congested, describing the disposition of flowers in an inflorescence.

Dentate : With sharp teeth perpendicular to the margin.

Deoxyribonucleic Acid (DNA) : Nucleic acid containing genetic instructions used for the proper functioning and development of all living organisms.

Deoxyribose : A type of sugar that is one component of DNA (deoxyribonucleic acid).

Depauperate : Starved or stunted, describing small plants or plant communities that are growing under unfavourable conditions.

Depressed : Flattened as if pressed down from the top or end, especially of 3.

Derepressible Enzyme : Enzyme produced in the absence of a specific inhibitory compound.

Dermal Tissue System : The protective covering of plants. In juvenile plants it is composed of a single layer of epidermal cells.

Desmid : Green algae that have conjugation of non-flagellate, amoeboid gametes.

Determinate : (1) of growth or branching, with a bud or flower terminating the main axis; (2) of an inflorescence (= anthotelic), with the inflorescence or parts of the inflorescence ending in a flower or an aborted but distinctly floral bud, e.g. panicle, thyrsoid, dichasium, monochasium.

Determinate Growth : A type of growth in which the organism stops growing after reaching the optimum size for that species.

Detritivore : Organisms that feed on dead, decomposed or organic waste.

Detritus : Organic matter that is either freshly dead or partially decomposed.

Development : Changes pertaining to the growth and differentiation of plant cells into various tissues and organs.

Developmental Response : The development of morphological and physiological qualities of an organism in response to prolonged or changing environmental conditions.

Dew Point : The temperature to which air must be cooled to bring about the condensation of water vapour.

Dextrorse : Turned to the right or spirally arranged to the right.

Di : Prefix meaning two or twice.

Diadelphous : Having the stamens united by their filaments into two groups, as in Fabaceae subfamily Faboideae.

Diapause : A period of inactive hormonal development as a response to unfavourable environmental conditions. This is a temporary phase.

Diaphram : A partition between two chambers.

Diatom : Unicellular, microscopic, freshwater or marine algae belonging to phylum Chrysophyta, which contain two silica shells fitting together like parts of a petri dish.

Diatomaceous Earth : Siliceous geological deposits made up of diatom frustules.

Diazotroph : Organism capable of using dinitrogen as its sole nitrogen source.

Dibbles : A tool used to plant bulbs and other small plant by poking holes in the soil.

Dichasium : A cyme in which branches appear in regular opposite pairs.

Dichotomous **:** Divided into two equal forks, e.g. of the branching pattern of stems or veins.

Dichotomous Branching **:** Fork resulting into two somewhat equal branches, as in the case of leaf veins or secretory ducts.

Dicot **:** One of two main division of flowering plants (the other being monocots), characterised by having two cotyledons (seed leaves); examples include most fruiting and flowering trees, and most annual and perennial flowering plants.

Dicotyledons **:** A major group of angiosperms (flowering plants) characterised by the embryo usually having two (rarely more) cotyledons (seed leaves).

Dictyosome **:** Organelle comprising disc-shaped, mostly branching hollow tubes, which accumulate and pack substances required for the synthesis of various materials in the cell.

Dictyostele **:** A complex stele with large overlapping leaf-gaps, in section composed of many meristeles (viz.).

Didymous **:** Twinned, the two parts similar and attached by a short portion of their inner surface.

Didynamous **:** With two pairs of stamens of unequal length.

Dieback **:** As a result of a lack of water, nutrients, or other stresses the stems of plants die.

Differential Medium **:** A medium with certain indicators, which helps distinguish between different chemical reactions during growth of organisms on it.

Differentially Permeable Membrane **:** Membrane permitting the diffusion of various substances at different rates.

Differentiation **:** Conversion of relatively unspecialised cell to a better specialised cell.

Diffuse **:** Of opening or straggling form; spreading and much-branched.

Diffuse Coevolution **:** Evolution of a species depending upon the evolution of some other species, which itself may be evolving depending on some other factors. Coevolution is basically, the evolution of groups depending on each other together, in order to survive.

Diffuse Competition : The weak interactions between species that are ecologically and distantly related.

Diffuse Growth : Generalised growth, not localised at apex or base.

Diffused Air Aeration : A diffused air activated sludge plant takes air, compresses it and discharges it with force, below the surface of water.

Diffusion : Haphazard movement of molecules from regions of high concentration to regions of lower concentration, leading to uniform distribution and levelling of the different concentration areas.

Digestion : Conversion of insoluble, complex substances into soluble, simpler substances under the control of enzymes.

Digitate : Radiating from a common point, having a fingered shape, i.e. a shape like an open hand.

Dihybrid Cross : Cross involving heterozygous parents with two different gene pairs.

Dikaryon : When two nuclei are present in the same hyphal compartment (they maybe homokaryon or heterokaryon), it is known as dikaryon.

Dikaryotic : Presence of two nuclei in a cell.

Dilution Plate Count Method : A method of estimating the number of viable microorganisms in a sample.

Dimidiate : Of a leaflet or pinnule with the lamina lacking or nearly lacking on the basiscopic side.

Dimorphic : Occurring in two different forms.

Dimorphism : The occurrence or existence of two forms within the same species bearing distinct structure, features, coloration, etc.

Dinitrogen Fixation : Conversion of molecular dinitrogen into ammonia and other organic combinations useful in other biological processes.

Dioecious : Plants featuring unisexual cones or flowers, with the male cones and flowers belonging to certain plants and the female cones and flowers confining to certain other plants of the same species.

Diploid : A full set of genetic material consisting of paired chromosomes, one from each parental set. Most animal cells except the gametes have a diploid set of chromosomes. The diploid human genome has 46 chromosomes.

Direct Count : Using direct microscopic examination to determine the number of microorganisms present in a given mass of soil.

Directed Evolution : A laboratory process used on isolated molecules or microbes to cause mutations and identify subsequent adaptations to novel environments.

Directed Mutagenesis : Alteration of DNA at a specific site and its re-insertion into an organism to study any effects of the change.

Directed Sequencing : Successively sequencing DNA from adjacent stretches of chromosome.

Disc : An outgrowth of tissue from the receptacle in the form of a ring or plate, sometimes divided into lobes or separate bodies, occurring between whorls of floral parts; generally glandular.

Disc Florets (Disc Flowers) : Usually actinomorphic flowers produced in the central part of the head and with a tubular corolla with more or less equal lobes, as in most Asteraceae.

Disciform : Having a flowering head that contains both filiform and disk flowers, referring to members of the *Asteraceae.*

Discoid : Having only disk flowers, referring to flower heads in the *Asteraceae.*

Discolorous : Of different colour, eg. when two sides of the leaf are different colours, also variegated.

Discrete : Clearly separate from each other, not united.

Disease-associated Genes : Alleles carrying particular DNA sequences associated with the presence of disease.

Disinfectant : An agent that kills microorganisms.

Disjunct : Separated from the main distribution of the population.

Disjunct-opposite : A variant of opposite and decussate leaf arrangement in which during development the leaves of a pair become separated on the axis owing to elongation of the nodal region, often giving the appearance of 'alternate' leaf arrangement but distinguished by decussate, not spiral, sequence.

Disk : The central portion of composite flowers, made up of a cluster of disk flowers.

Dissected : Deeply cut or divided into numerous lobes or divisions.

Distal : The end opposite the point of attachment, away from the axis.

Distant : Far from each other or from something else.

Distichous (2-ranked) : Arranged in two rows on opposite sides of a stem and in the same plane.

Distinct : Having separate, like parts, those not at all joined to each other, often describing the petals on a flower.

Disturbed : Referring to habitats that have been impacted by the actions of people.

Diuretic : Substances that increase the urine flow.

Diurnal : Having daily periodic cycle, e.g. endogenous circadian rhythm.

Divaricate : Widely diverging or spreading apart.

Divergent : Spreading away from one another, usually at a wide angle.

Divergent Speciation : Emergence of a new species from a part of existent species, with the remnant species continuing as the original species itself, or else transforming into a new species.

Divided : Cut deeply, nearly or completely to the midrib.

Division : Refers to the process of propagating perennials. Generally after 3 years a perennial is a well developed and can be divided by digging it up and splitting root ball. Often times the root ball will split naturally without much force into smaller plants that can be planted on their own.

DNA (Deoxyribonucleic Acid) : The molecule that encodes genetic information. DNA is a double-stranded molecule held together by weak bonds between base pairs of nucleotides. The four nucleotides in DNA contain the bases adenine (A), guanine (G), cytosine (C), and thymine (T). In nature, base pairs form only between A and T and between G and C; thus the base sequence of each single strand can be deduced from that of its partner.

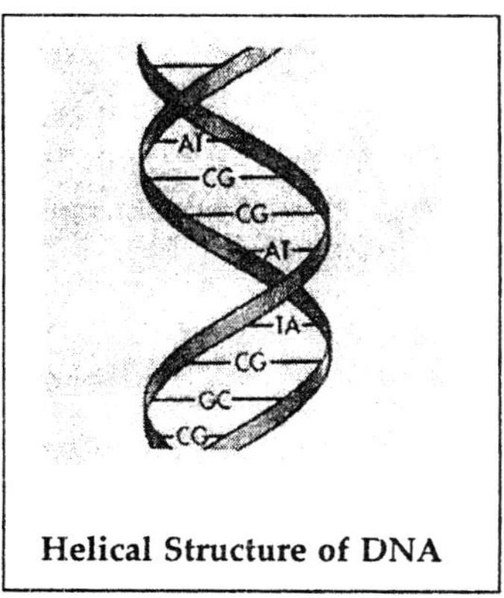

Helical Structure of DNA

DNA Bank : A service that stores DNA extracted from blood samples or other human tissue.

DNA Fingerprinting : Techniques by which possible differences between different DNA samples can be assessed.

DNA Repair Genes : Genes encoding proteins that correct errors in DNA sequencing.

DNA Replication : The use of existing DNA as a template for the synthesis of new DNA strands. In humans and other eukaryotes, replication occurs in the cell nucleus.

DNA Sequence : The relative order of base pairs, whether in a DNA fragment, gene, chromosome, or an entire genome.

Dodeca : Prefix meaning twelve.

Dolipore Septum : Specialised cross-wall that separates hypha of fungi belonging to the genus Basidiomycota.

Domain : A discrete portion of a protein with its own function. The combination of domains in a single protein determines its overall function.

Domatia : Small structures on the lower surface of some leaves, usually consisting of depressions, partly enclosed by leaf tissue or hairs, usually located in the axils of the primary and secondary veins.

Dominance : Phenomenon in which one allele of a gene masks the phenotypic expression of another allele of a gene. The allele masking the other allele is called dominant allele.

Dominant : An allele that is almost always expressed, even if only one copy is present.

Dommatia : Pits or pockets, sometimes with tufts of hair in the axils of nerves on the lower surfaces of leaves.

Dormancy : The phase of temporary growth cessation in plants, under harsh environmental situations, wherein the regular conditions required for growth cannot be met.

Dorsal : Relating to the back of an organ, i.e. the surface of a lateral organ facing away from the axis.

Dorsifixed : Attached at the back.

Dorsiventral : Term describing leaves in which the upper and lower surfaces differ from each other in texture, presence of hairs, stomates etc.

Double Flowers : A form of flower in which there are several rows each with many petals. An example is the rose.

Double Fusion : Phenomenon in which one sperm fertilises an egg to form a zygote, while another sperm fertilises the central cell nuclei (polar nuclei) forming a primary endosperm nucleus.

Double Helix : The twisted-ladder shape that two linear strands of DNA assume when complementary nucleotides on opposing strands bond together.

Doubling Time : The time needed for a certain population to double in number.

Draft Sequence : The sequence generated by the HGP as of June 2000 that, while incomplete, offers a virtual road map to an estimated 95% of all human genes. Draft sequence data are mostly in the form of 10,000 base pair-sized fragments whose approximate chromosomal locations are known.

Drainage : The act of draining water through the soil. Sandy soil has better drainage (loses water faster) than does clay soil (loses water slower).

Drip Emitters : Part used in drip irrigation systems in order to deliver water as a drip.

Drip Irrigation : A system of irrigation using small diameter hoses to deliver water at slow rates (i.e. drips). Useful in watering small areas.

Drooping : Erect or spreading at the base, then bending downwards.

Drupaceous : Term describing a fruit which is a drupe or drupe-like.

Drupe : A fleshy indehiscent fruit enclosing a nut or hard stone containing generally a single seed such as a peach or cherry.

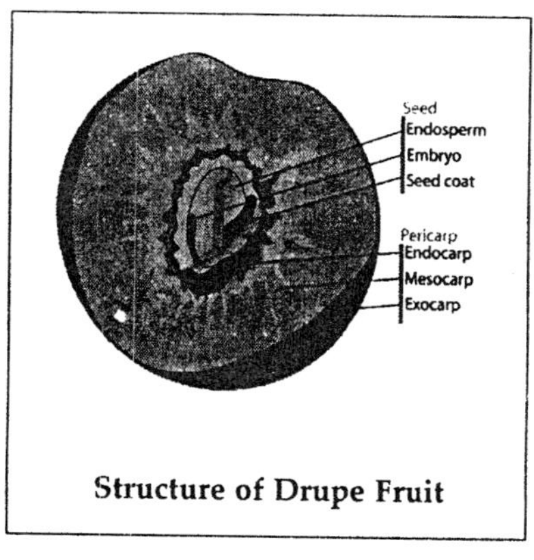

Structure of Drupe Fruit

Drupelet : One drupe of a mature fruit composed of a cluster of small drupes, as in blackberries, the 'seed' being a pyrene.

Dry Sclerophyll Forest : An open forest in which scleromorphic (hard-leaved) shrubs form a layer below the trees (usually species of eucalypts).

E

Early Wood : Wood formed during the early part of the growing season, characterised by large, thin walled cells. It features large number of vessels in angiosperms and in gymnosperms, it features wide tracheids.

Eccentric : Off-centre, not positioned directly on the central axis.

Ecological Isolation : It refers to the situation where closely related (sometimes virtually indistinguishable) species live in the same territory, but slight differences in their niche causes them to reproduce in isolation of the others.

Ecological Release : It refers to the progression in which a species expands its living habitat as well as the resources it utilises into areas that have a lower density of species in terms of diversity.

Ecology : Branch of biology involving study of interactions of organism with the environment and with each other.

Ecomorphology : It is the study of the relation between an individual's ecological role and its form and structural adaptations.

Ecosystem : System involving the interactions of living organisms with each other as well as with the non-living environment.

Ecosystem Approach : This is a method of resource management that acknowledges that the different components of an ecosystem (structure, function, and species composition) are inter–linked, and this factor must be taken into consideration while restoring and protecting the ecosystem's natural balance.

Ecotone : Transition zone between two adjoining communities.

Ecotype : Those individuals adapted to a specific environment or set of conditions.

Ecto : Prefix meaning 'outside'.

Ectomycorrhiza : One type of mycorrhizal association, wherein the fungi do not invade the cell membrane, instead invade the root cortex cells.

Ectoparasite : An organism, such as a tick, that latches itself to the surface of its host, in order to survive.

Ectophloic Siphonostele : A cylindrical stele with a parenchymatous central strand and phloem around the outside only (= medullated protostele).

Ectothermy : Refers to an organism's ability to maintain its body temperature by availing heat from the environment, either by absorbing radiation or through conduction.

Effective Population Size : The average size of a population expressed in terms of individuals assumed to contribute genes equally to the next generation; generally smaller than the actual size of the population, depending on the variation in the reproductive success among individuals.

Egestion : Removal of undigested food material.

Egg : Non motile female gamete.

Elater : A cell or part of a cell which assists in dispersing spores. The elaters change shape as they lose or acquire water, and they will then push against surrounding spores.

Electron : A component of an atom containing a negative charge which orbits around the nucleus.

Electron Transport Chain : A system located in the inner membrane of the mitichondrion and in the thylakoid membranes of chloroplasts that produces energy in the form of ATP via a slide of electrons down a gradient.

Electrophoresis : A method of separating large molecules (such as DNA fragments or proteins) from a mixture of similar molecules. An electric current is passed through a medium containing the mixture, and each kind of molecule travels through the medium at a different rate, depending on its electrical charge and size. Agarose and acrylamide gels are the media commonly used for electrophoresis of proteins and nucleic acids.

Electroporation : A process using high-voltage current to make cell membranes permeable to allow the introduction of new DNA; commonly used in recombinant DNA technology.

Ellipsoid : Of a solid body with an elliptic section or outline.

Elliptic : Of a plane with the shape of an ellipse, longer than wide and rounded at both ends, the widest part near the middle.

Elliptical : Wider in the middle and narrowing equally towards the ends; in the shape of an ellipse.

Elongate : Stretched out, many times longer than broad.

Emarginate : Cut out or hollowed; notched, especially at the ends.

Embossed : With a small central nodule.

Embryo : An animal or plant that is in its first stage of development and is usually still contained within the seed, egg, or uterus.

Embryonic Stem (ES) Cells : An embryonic cell that can replicate indefinitely, transform into other types of cells, and serve as a continuous source of new cells.

Embryophyte : Synonym for the Plantae, as here defined. It includes all green photosynthetic organisms which begin the development of the sporophyte generation within the archegonium.

Emergent : Of a plant, (1) rising above the surrounding plants, e.g. of a tree above the rainforest canopy; (2) rising above the surface of the water.

Emitter Line : A part used in drip irrigation systems to deliver or emit water to plants along the hose line.

Enation : Tiny green leaf like structures growing on the stems of whisk ferns and do not have vascular tissue.

Endangered Species : The entire population of organisms (plant or animal) that face extinction due to a steady reduction of their numbers. This may be the outcome of environmental changes, loss of habitat or predation.

Endemic : Having a natural distribution confined to a particular geographic region..

Endo : Prefix meaning 'inside'.

Endocarp : The inner layer of the pericarp, which is the wall of the ripened ovary or fruit.

Endocytosis : Absorption of solid or liquid material into a cell by means of invagination of the plasma membrane to surround the material and pinching shut to form a vacuole or vesicle around it.

Endodermis : Literally "inner skin", this is a layer of cells which surrounds the central core of vascular tissue, and which helps to regulate the flow of water and dissolved substances.

Endoenzyme : Enzyme that acts along the internal portion of a polymer.

Endonuclease : The endoenzyme responsible for breaking the phosphodiester bonds in a nucleic acid molecule.

Endoparasite : This type of organism or parasite (such as tapeworm) exists and feeds inside the bloodstream or tissue of its host.

Endophyte : An organism, which maybe parasitic or symbiotic, with a plant that is grown within.

Endoplasmic Reticulum : Complex system of narrow tubes and sheets forming a network in the cell's cytoplasm. It divides the cytoplasm into various compartments. The endoplasmic reticulum may or may not have ribosome attached to them.

Endosperm : Nutritive material derived from the embryo sac in seed plant ovules. It furnishes the developing embryo and seedling with nutrition.

Endospore : A cell which is formed by certain gram-positive bacteria in unfavourable conditions. An endospore is extremely resistant to heat and other harmful agents.

Endosymbiont Hypothesis : According to this hypothesis, mitochondria and plastids were free-living bacteria, which got incorporated into the cells.

Endothermic : The ability of an organism to constantly maintain its body temperature, usually keeping itself warm, irrespective of the external conditions.

Endothermy : The ability of an organism to maintain its body temperature, by generating heat metabolically.

Enhanced Rhizosphere Degradation : Enhanced activity of microorganisms involved with biodegradation of contaminants near plant roots which is brought about by compounds exuded by the plant roots.

Enrichment Culture : Technique wherein environmental conditions are altered to aid the growth of a specific organism or group of organisms.

Enteric Bacteria : These are bacteria present in the intestinal tract of humans and other animals. They maybe physiologic or pathologic.

Entire : Describing a leaf that has a continuous, unbroken margin with no teeth or lobes.

Entomophily : Seed plants which are pollinated by insects are said to be entomophilous.

Enzyme : A protein that acts as a catalyst, speeding the rate at which a biochemical reaction proceeds but not altering the direction or nature of the reaction.

Ephemeral : Describes a plant or flower that lasts for only a short time or blooms only occasionaly when conditions are right.

Epi : Prefix meaning 'on or above'.

Epicalyx : A whorl of bracts just below the flower, resembling an extra calyx.

Epicarp : The outermost layer of the pericarp.

Epicormic : Term describing buds, shoots or flowers borne on the old wood of trees, often applied to shoots arising from dormant buds after injury or fire, as in eucalypts.

Epicotyl : Portion of the seedling above the cotyledon's attachment point.

Epidermis : The outer most layer of cells in an organism. In plants the epidermis consists of the dermal tissue system.

Epigynous : Of floral parts, especially stamens, inserted on or above the ovary, and arising from tissue that is fused to the ovary wall.

Epimatium : The ovule-bearing scale in some conifers, as in Podocarpus species.

Epiphyllous : Growing on leaves, as plantlets on the leaves in some Crassulaceae.

Epiphyte : A plant which grows upon another plant. The epiphyte does not "eat" the plant on which it grows, but merely uses the plant for structural support, or as a way to get off the ground and into the canopy environment.

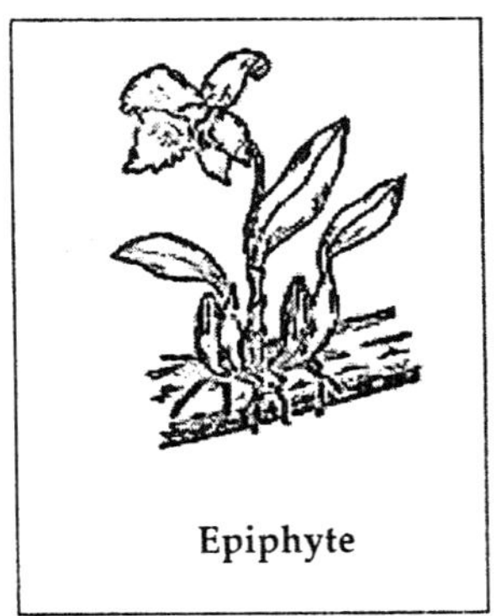

Epiphyte

Episome : An extrachromosomal replicating genetic element found in certain bacteria.

Epistasis : One gene interfere with or prevents the expression of another gene located at a different locus.

Epitope : An antigenic determinant of known structure. It is the region of the antigen to which the variable region of the antibody binds.

Equitant : Of leaves, folded longitudinally with the two inner surfaces (representing the upper leaf surface) fused except towards the base where it clasps another leaf on the opposite side of the stem; one margin represents the leaf keel and the lamina is vertically orientated; as in Iris. Leaf arrangement; cross section through equitant arrangement.

Erect : (1) upright, e.g. of a shrub; (2) perpendicular to a surface, e.g. of hairs.

Erecto-patent : Between spreading and erect.

Eremean : Pertaining to regions of low, irregular rainfall.

Ericoid : Of leaves, small and sharply pointed like those of the heaths.

Ericoid Mycorrhizae : The type of mycorrhizae found in Ericales plants. These hyphae are capable of penetrating cortical cells.

Erose : Having an irregular margin as if it has been gnawed.

Escapee : A plant that has escaped from cultivation and now reproduces on its own.

Escent : Suffix meaning 'inclined to be' or 'becoming', eg. accrescent, coalescent.

Escherichia Coli : Common bacterium that has been studied intensively by geneticists because of its small genome size, normal lack of pathogenicity, and ease of growth in the laboratory.

Espalier : A plant, typically a tree or shrub, that has been trained to grow in a flattened pattern with the help or wires, trellis, or a fence.

Essential Element : Elements which are essential for normal development, growth and reproduction of plants. Nitrogen, calcium, magnesium, phosphorus, etc. are examples of some essential elements.

Estaurine : Pertaining to estuaries or river mouths, usually brackish conditions.

Estivation : Similar to hibernation, it is a period of inactivity that the animal goes into during a dry hot season.

Estuaries : Water bodies located at river ends. They are subjected to tidal fluctuations.

Ethylene : A plant hormone most note able involved in fruit ripening, but also involved in aging, growth inhibition, and leaf abscission. Ethylene is a gaseous hormone.

Etiolation : Term referring to a condition involving poor leaf development, long inter nodes, pale and weak appearance of the plant due to deprivation of sunlight.

Eubacteria : A genus of bacteria belonging to the family Propionibacteriaceae, found as saprophytes in soil and water.

Eugenics : The study of improving a species by artificial selection; usually refers to the selective breeding of humans.

Eukaryote : Cell or organism with membrane-bound, structurally discrete nucleus and other well-developed subcellular compartments. Eukaryotes include all organisms except viruses, bacteria, and blue green algae.

Eukaryotic : Cells comprising nucleus, chromosomes and distinct membrane bound organelles.

Eusporanagiate : Of primative ferns with sporangial walls more than one cell thick originating from several cells.

Eustele : When a plant's vascular tissue develops in discrete bundles, it is said to have a eustele.

Eutrophication : Process of nutrient accumulation in the water bodies resulting in its gradual nutrient enrichment. This entails to increase in the growth of algae and various other organisms.

Evanescent : Fleeting, lasting for only a short time.

Even-pinnate : A pinnately-compound leaf ending in a pair of leaflets.

Evergreen : Plants that maintain their foliage year round as opposed to deciduous plants.

Evolution : At the most basic level, evolution is change that takes place over time. In reference to lifeforms, evolution is the genetic changes observed amongst the population of organisms from generation to generation.

Ex : Prefix meaning 'without or lacking' or meaning 'outwards'.

Excentric : Not centrally placed, without a fixed centre.

Excrescence : Outgrowth from the surface.

Excurrent : The stem or trunk continuing to the top of a tree; running out, as a vein projecting beyond the margin of a leaf.

Exfoliate : To come away in scales or flakes.

Exfoliating : Peeling off or "shedding" in thin layers.

Exine : The outer coat of a pollen grain or spore.

Exocarp : The outer layer of the pericarp of a fruit.

Exoenzyme : An enzyme which acts outside the cell that secretes it.

Exogenous DNA : DNA originating outside an organism that has been introducted into the organism.

Exons : The region of a split DNA that codes for RNA.

Exonuclease : An enzyme that cleaves nucleotides sequentially from free ends of a linear nucleic acid substrate.

Exotic : Introduced from outside the area concerned, in the case of N.S.W. usually from overseas.

Explant : Severed portions of the plant; example : leaf or stem tissue that are utilised for tissue culture.

Expressed Sequence Tag (EST) : A short strand of DNA that is a part of a cDNA molecule and can act as identifier of a gene. Used in locating and mapping genes.

Exserted : Projecting beyond the surrounding objects, e.g. of stamens protruding beyond the perianth, or of valves projecting beyond the rim of a capsular fruit.

Extant : Still surviving, not completely extinct.

Extirpated : Destroyed or no longer surviving in the area being referred to, but may survive outside of that area.

Extrafloral : Not within the flower, usually applied to nectariferous glands, e.g. as those on the petiole in some Croton species and on the phyllodes of some wattles.

Extranuclear DNA : DNA located outside the nucleus, as seen in mitochondria and plastids.

Extrorse : Turned or opening outward away from the axis.

Exudate : A liquid, resinous or gelatinous substance secreted by organs or parts of the plant, or yielded when the plant is damaged.

Eyespot : Tiny reddish sensory organ, which is sensitive to light. It is found within motile unicellular organisms.

F

F1 Generation : The first generation of a genetic cross.

F2 Generation : The offspring resulting from the interbreeding of individuals in the F1 generation.

Facultative : Exhibiting some ability or function under some environmental conditions but not under others.

Facultative Aerobe : Organisms that use oxygen when available, however, can even live without it.

Facultative Organism : An organism which is able to adjust to a particular circumstance or has the ability to take up different roles in a process.

False Indusium : A covering of the sorus formed from the reflexed margin of the lamina.

False Veins : Zones of epidermal cells of similar appearance to veins but not associated with vascular tissue, eg. in some Davallia and Hymenophyllaceae.

Family : It is a term of classification of living things, in which this group falls below an order. It is further divided into one or more genera. The ranks start with life, followed by domain, kingdom, phylum, class, order, family, genus and species.

Fang : Fangs are long, pointed teeth located in the front of the mouth. In mammals, these fangs are called canine teeth and are used for tearing flesh. In snakes, they are used to inject venom into the victim.

Farina : A mealy or flour-like covering; hence farinose, farinaceus.

Farinose : Covered with a mealy or whitish powdery substance.

Fascicle : A small cluster or bundle, a fairly common leaf arrangement.

Fascicled : Arranged in bundles or clusters, e.g. leaves.

Fastigiate : Clustered, parallel and erect, having a broom-like appearance.

Fauna : All the animal life that exists in a particular area during a specific period of time.

Feces : Indigestible waste products expelled from an organisms digestive tract. Usually referred to as stool.

Fecundity : In a general sense, it refers to an organisms ability to reproduce. In biology, it refers to a females potential capacity to reproduce, based on the number of gametes (eggs), seed set or asexual propagules.

Feedback Initiation : Inhibition by an end product of the biosynthetic pathway involved in its synthesis.

Feldmark : High altitude plant community characterised by scattered, dwarf prostrate plants with a mat or cushion habit.

Felted : Matted with very short interlocked hairs, having the appearance or texture of felt.

Femur : In vertebrates having four limbs the femur is the upper bone of the hind limb. In insects, it is the third segment in the leg.

Fenestrate : With small slits or areas thinned so as to be translucent.

Fermentation : Type of respiration involving the process of glycolysis, wherein lactic acid or ethyl alcohol are formed as an end product.

Fertile : Capable of reproducing itself; also used of portions of a plant or organ producing reproductive structures.

Fertilisation : The union of female and male gametes.

Fertiliser : Any organic or inorganic material added to the soil to enhance the growth of plants.

Fetus : A fetus is a developing organism, which has moved beyond the embryonic stage, but is yet to be born.

Fiber : Elongated and thickened cell found in xylem tissue. It strengthens and supports the surrounding cells.

Fibrillose : Finely fibrous furnished with threadlike structures.

Fibrous Root : Root system consisting of many small roots, often forming a mat which spreads out vertically and laterally in the soil. Fibrous roots do not usually penetrate very deeply in the soil. This type of system is often characteristic of monocots.

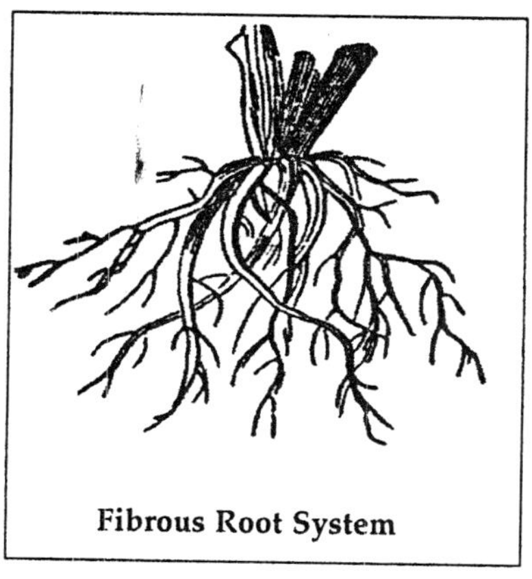

Fibrous Root System

Field Capacity : Content of water remaining in the soil after being saturated with water.

Filament : A threadlike structure; that part of the stamen which supports the anther.

Filamentous : In the form of very long rods, mostly seen in bacteria. Seen as branching strands in fungi.

Filamentous Body : Usually green algae cells exhibit this kind of body, wherein the cells are held firmly by a middle lamella when they divide transversely.

Filial Generation (F1, F2) : Each generation of offspring in a breeding program, designated F1, F2, etc.

Filiform : (1) threadlike; (2) a type of flower in the Asteraceae which is pistillate and has a very slender, tubular corolla.

Filly : A female horse that is four years or younger in age.

Filter Feeder : Organisms that feed by sieving water for food particles, with the help of special filtering structures in their mouths. Clams, sponges, krill and baleen whales use this method.

Fimbria : Short filamentous structure present on a bacterial cell, involved with adhesion of the bacteria to other surfaces it comes in contact with.

Fimbriate : Fringed, the edge bordered by slender processes (dimin. fimbrillate).

Fingerprinting : In genetics, the identification of multiple specific alleles on a person's DNA to produce a unique identifier for that person.

Finished DNA Sequence : High-quality, low error, gap-free DNA sequence of the human genome. Achieving this ultimate 2003 HGP goal requires additional sequencing to close gaps, reduce ambiguities, and allow for only a single error every 10,000 bases, the agreed-upon standard for HGP finished sequence.

First Filial Generation : Progeny of an experimental cross between two parent species.

Fission : Cell division of bacterial and other related organisms that results into two new cells.

Fistular : Cylindrical and hollow like a pipe.

Fitness : The measure of an individual's genetic contribution to the next generation's gene pool.

Flabellate : Fan-shaped, as in a fan-shaped structure.

Flagellum : Multicellular organisms produce threadlike structures, which protrude from the motile cells and assist mainly in locomotion.

Flexuose or Flexuous : With curves or bends, somewhat zigzagged.

Flexuous : Zigzag or curved alternately in opposite directions.

Flora : (1) the assemblage of plant taxa of an area; (2) a book dealing systematically with the plants of an area.

Floret : (1) a small flower, one of a spikelet or densecluster, as in Asteraceae; (2) a grass flower, together with the lemma and palea that enclose it.

Floricane : The second-year flowering and fruiting cane or shoot of *Rubus*.

Floridean Starch : Branched glucan polymer, storage produce of Rhodophyta present as granules outside the chloroplasts.

Flow Cytometry : Analysis of biological material by detection of the light-absorbing or fluorescing properties of cells or subcellular fractions (i.e., chromosomes) passing in a narrow stream through a laser beam. An absorbance or fluorescence profile of the sample is produced. Automated sorting devices, used to fractionate samples, sort successive droplets of the analysed stream into different fractions depending on the fluorescence emitted by each droplet.

Flow karyotyping : Use of flow cytometry to analyse and separate chromosomes according to their DNA content.

Flower : The sexual reproductive structure of angiosperms, typically consisting of an axis bearing perianth parts, androecium and gynoecium.

Fluid Mosaic Model : A plasma membrane model according to which, the proteins are embedded in the lipids throughout the membrane which gives a mosaic appearance to it. The proteins change their position, hence, it is called a 'fluid' membrane.

Fluorescence in Situ Hybridisation (FISH) : A physical mapping approach that uses fluorescein tags to detect hybridisation of probes with metaphase chromosomes and with the less-condensed somatic interphase chromatin.

Fluorescent Antibody : This is a laboratory test that is done, wherein antibodies are tagged with fluorescent dye to detect the presence of microorganisms.

Flush : A period of rapid vegetative growth, often involving the expansion of non-green or pale green leaves at the soot extremeties.

Fluted : Regularly marked by alternating rounded ridges and groovelike depressions.

Foal : A male or female horse that is up to six months old.

Foetid : With a stinking odour; smelling offensively.

Foliar : Pertaining to the leaves or leaflike parts.

Foliate : Leaved, clothed in leaves; also as a suffix, eg. bifoliate.

Foliolate : An adjective used with a number prefix to indicate the number of leaflets forming a compound leaf.

Foliose : Bearing numerous crowded leaves.

Follicle : A dry, many-seeded fruit derived composed of a single carpel l and opening along one side only like a milkweed pod.

Food Chain : The food chain is the transfer of life giving energy from one organism to another, which is compatible to receive the same form of energy that was passed on, when the organism perished.

Simple Food Chain

Food Web : A food web comprises a set of interconnected food chains which exist within an ecosystem.

Foot : Attached to the gametophyte, the foot is the basal portion of the embryo of bryophytes and absorbs food from the gametophyte.

Forb : A non-grasslike herbaceous plant.

Forensics : The use of DNA for identification. Some examples of DNA use are to establish paternity in child support cases; establish the presence of a suspect at a crime scene, and identify accident victims.

Forest : A plant community dominated by long-boled trees in close proximity.

Form (forma) : The smallest taxonomic category, generally used for variations occurring among individuals of any population.

Founder : Individuals who are the first to establish a population in a new environment or habitat.

Founder Effect : In terms of describing the genetic outcome of a new population being established by a very small number of individuals, from a larger population. The founder effect refers to the loss of genetic variation.

Fountain Growth : Multiaxial growth usually of many filaments in which cell divisions occur.

Fragmentation : Production of new individuals from fragments of original individual.

Fraternal Twin : Siblings born at the same time as the result of fertilisation of two ova by two sperm. They share the same genetic relationship to each other as any other siblings.

Free : Individually arising or inserted, not united, nor fused, adherent, adnate, connate etc.

Free-central : Of placentation, with the placenta along the central axis in a compound ovary without septa.

Freely Permeable Membrane : Membranes permitting all kinds of substances to pass through it.

Frond : Usually used for a fern leaf, however, occasionally it is also used to denote palm leaves.

Frugivorous : Fruit-eating. A frugivore is any organism whose preferred food type is fruits.

Fruit : In flowering plants, the structure which encloses the seeds. True fruits develop from the ovary wall, such as bananas and tomatoes, though not all fruits are edible, such as the dry pods of milkweed or the winged fruits of the maple.

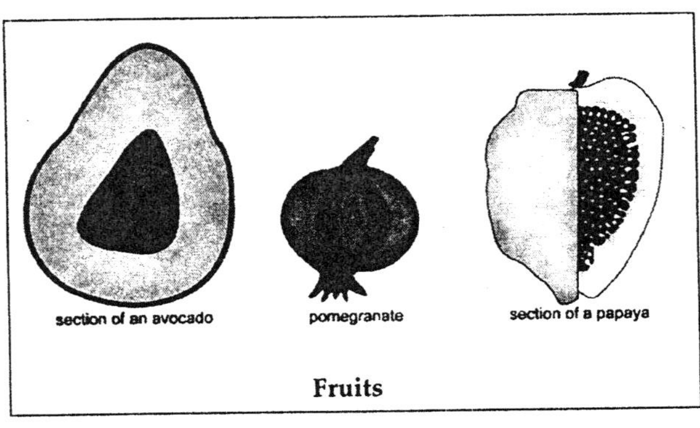

Fruits

Frustule : Siliceous wall and protoplasm seen in diatoms.

Frutescent : Shrubby or bushy in the sense of being woody.

Fruticose : Shrub-like; hence frutescent, becoming shrub-like.

Fucosan Vesicle : Small, refractive particle or physode in cell of Phaeophycea, containing tannin-like compounds.

Full Gene Sequence : The complete order of bases in a gene. This order determines which protein a gene will produce.

Fulvic Acid : The yellow organic material that remains behind after removal of humic acid by the process of acidification.

Fulvous : Yellow, tawny, dull yellow with a mixture of grey or brown.

Functional Genomics : The study of genes, their resulting proteins, and the role played by the proteins the body's biochemical processes.

Functional Response : The relationship between prey and predator, or deviation in the rate of exploitation of prey by an individual predator due to change in prey density.

Fungistasis : Suppression of growth of new fungal cells, due to excessive competition for nutrients, or due to the presence of excessive inhibitory compounds in the soil.

Fungus : Eukaryotic heterotrophic organisms that live as saprophytes or parasites. This group includes mushrooms, yeast and molds. They have a rigid cell wall.

Funnelform : Gradually widening upwards, as in the flowers of morning glory.

Furcate : Forked, also as bifurcate, trifurcate etc.

Furcellaran : Sulfated polysaccharide gel (like carrageenan) found in cell wall of red alga *Furcellaria.*

Furfuraceous : Scurfy, branlike, flaky.

Furfuraceus : Scurfy, provided with soft scales.

Furrowed : With +/- parellel longitudinal grooves or channels.

Fuscous : Dusky, blackish, of a brownish or greyish brown colour.

Fused : Joined and growing together.

Fusiform : A spindle- or cigar-shaped, a solid swollen in the middle and tapering at both ends.

Fusion Cell : Cell produced by fertilisation.

G

Galea : In Orchidaceae, a perianth segment or group of perianth segments shaped like a helmet.

Gall : An abnormal growth on a plant that is caused by insects.

Gametangium : Cell in which gametes are produced.

Gamete : The male and female sexual cells that combine at fertilisation to form the zygote; in pteridohytes produced on the prothallus (gametophyte) by the antheridia (male) and archegonia (female).

Gametophore : Leafy stalk on which the gametangium(sex organs) is borne.

Gametophyte : The sexual or haploid stage in the life cycle of crytogams producing the ova and sperm; in pteridophytes it is a separate, inconspicuous and short-lived generation.

Gamma Diversity : This term refers to the measure of biodiversity, which means the total species richness within an area.

Gamophyllous : Having the bases of opposite leaves fused around the stem.

Gaping : The open width of space, created by forcefully opening the jaws or mandibles of a vertebrate.

Gas Vacuole : A sub : Cellular organelle, found only in prokaryotes, which are gas filled vesicles.

GC-rich Area : Many DNA sequences carry long stretches of repeated G and C which often indicate a gene-rich region.

Gelatinous : Of a slimy, clear sticky nature, water soluble, like gelatine.

Gelding : A male horse that has been spayed.

Gemma : Cluster of cells that get detached from parent body and possess the ability to develop into a completely new organism or plant. Seen in liverworts and mosses.

Gene : The fundamental physical and functional unit of heredity. A gene is an ordered sequence of nucleotides located in a particular position on a particular chromosome that encodes a specific functional product (i.e., a protein or RNA molecule).

Gene Amplification : Repeated copying of a piece of DNA; a characteristic of tumour cells.

Gene Bank : It is a way of preserving plants and seeds for their germ plasm.

Gene Chip Technology : Development of cDNA micro–arrays from a large number of genes. Used to monitor and measure changes in gene expression for each gene represented on the chip.

Gene Cloning : Isolation of a desired gene from an organism and its replication in large amounts. It is used extensively in DNA research.

Gene Expression : The process by which a gene's coded information is converted into the structures present and operating in the cell. Expressed genes include those that are transcribed into mRNA and then translated into protein and those that are transcribed into RNA but not translated into protein (e.g., transfer and ribosomal RNAs).

Gene Family : Group of closely related genes that make similar products.

Gene Mapping : Determination of the relative positions of genes on a DNA molecule (chromosome or plasmid) and of the distance, in linkage units or physical units, between them.

Gene Pool : Total number of all alleles in all the sex cells present in the individuals of a population.

Gene Prediction : Predictions of possible genes made by a computer program based on how well a stretch of DNA sequence matches known gene sequences.

Gene Probe : A strand of nucleic acid which can be labelled and hybridised to a complementary molecule from a mixture of other nucleic acids. It is helpful in DNA sequencing.

Gene Product : The biochemical material, either RNA or protein, resulting from expression of a gene. The amount of gene product is used to measure how active a gene is; abnormal amounts can be correlated with disease-causing alleles.

Gene Synthesiser : Machine producing specific DNA sequences.

Gene Therapy : An experimental procedure aimed at replacing, manipulating, or supplementing nonfunctional or misfunctioning genes with healthy genes.

Gene Transfer : Incorporation of new DNA into and organism's cells, usually by a vector such as a modified virus. Used in gene therapy.

Generalist : Any organism that can survive in wide ranging habitat.

Generation Time : The time required for a population to double in number.

Genetic Code : The sequence of nucleotides, coded in triplets (codons) along the mRNA, that determines the sequence of amino acids in protein synthesis. A gene's DNA sequence can be used to predict the mRNA sequence, and the genetic code can in turn be used to predict the amino acid sequence.

Genetic Counselling : Provides patients and their families with education and information about genetic-related conditions and helps them make informed decisions.

Genetic Discrimination : Prejudice against those who have or are likely to develop an inherited disorder.

Genetic Drift : Alteration in the genetic make-up of a particular population, which mostly takes place by chance alone.

Genetic Engineering : Introduction of genes from one DNA form into another, by artificial means is called genetic engineering.

Genetic Illness : Sickness, physical disability, or other disorder resulting from the inheritance of one or more deleterious alleles.

Genetic Marker : A gene or other identifiable portion of DNA whose inheritance can be followed.

Genetic Mosaic : An organism in which different cells contain different genetic sequence. This can be the result of a mutation during development or fusion of embryos at an early developmental stage.

Genetic Polymorphism : Difference in DNA sequence among individuals, groups, or population (e.g., genes for blue eyes versus brown eyes).

Genetic Predisposition : Susceptibility to a genetic disease. May or may not result in actual development of the disease.

Genetic Screening : Testing a group of people to identify individuals at high risk of having or passing on a specific genetic disorder.

Genetic Testing : Analysing an individual's genetic material to determine predisposition to a particular health condition or to confirm a diagnosis of genetic disease.

Genetics : Branch of biology involving the study of heredity, which deals with the differences and resemblances of organisms entailing from the interaction of their genes and the habitat.

Geniculate : Bent abruptly like a knee or a stove pipe.

Geniculum : Flexible portion of the thallus, such as the non-calcified segments between calcified intergenicula of articulated corallines.

Genome : All the genetic material in the chromosomes of a particular organism; its size is generally given as its total number of base pairs.

Genome Project : Research and technology-development effort aimed at mapping and sequencing the genome of human beings and certain model organisms.

Genomic Library : A collection of clones made from a set of randomly generated overlapping DNA fragments that represent the entire genome of an organism.

Genomics : The study of genes and their function.

Genotype : The genetic constitution of an organism, as distinguished from its physical appearance (its phenotype).

Genus : A taxonomic group of closely related species or a single species without close relatives; closely related genera are grouped into families. pl. genera.

Geophyte : A plant with an underground storage organ (e.g. corm, tuber, bulb or rhizome) and with annually renewed aerial shoots.

Germ Cell : Sperm and egg cells and their precursors. Germ cells are haploid and have only one set of chromosomes (23 in all), while all other cells have two copies (46 in all).

Germ Line : The continuation of a set of genetic information from one generation to the next.

Germ Line Gene Therapy : An experimental process of inserting genes into germ cells or fertilised eggs to cause a genetic change that can be passed on to offspring. May be used to alleviate effects associated with a genetic disease.

Germ Plasm : Aggregate of all genes of a species or organism groups.

Germination : Commencement or resumption of growth of a spore or seed.

Germ-line Mutation : Mutation occurring in the cells from which gametes are derived.

Gibberellin : Group of plant hormones possessing different effects on growth, which are mostly related to enhancement of stem elongation.

Gibbous : Somewhat swollen on one side, usually near the base, forming a pouch or sack.

Girdling : Phenomenon involving the discarding of a band of tissues which extend to the inner side of the vascular cambium on the woody plant stem.

Glabrescent : Becoming naked or devoid of covering at maturity.

Glabrous : Smooth in the sense of having no hairs; not pubescent.

Gland : A secreting surface or structure; usually small cellular organs which secrete oily or other products; sometimes sunk in, sometimes mounted on a stalk or at the tip of a hair or tooth. The name is also used for any protuberance or appendage resembling such a structure.

Gland Cell : Special, refractive cell in some red algae which may function in secretion or storage of compounds.

Glandular : Producing tiny globules of sticky or oily substance, e.g. of hairs, of a surface.

Glans : A dry dehiscent fruit borne in a cupule, such as the acorn.

Glaucescent : Slightly glaucous or becoming so.

Glaucous : Dull blue-green in colour, with a whitish bloom which can often be rubbed off; sometimes characteristic of young leaves, as in some eucalypts.

Gliding Movement : Movement of an organism when in contact with substratum.

Globose (Globular, Orbicular, Spherical) : A 3-dimensional shape, ball-shaped, more or less circular in outline.

Glochid : A barbed bristle, as in many Cactaceae.

Glochidium : A very dense cluster; hence glomerulate.

Glochids : Barbed bristles on cacti.

Glomerate : Crowded, congested or compactly clustered.

Glomerule : A small compact cluster, e.g. of flowers.

Glumaceous : Having the nature of or resembling a glume, tending to be chaffy or membranous in texture.

Glume : A bract in the inflorescence of some monocots; (1) one of the two bracts at the base of the grass spikelet; (2) also used in Cyperaceae and Restionaceae for the small bracts on the spikelet in which flower is subtended.

Glutinous : With a sticky covering.

Glycocalyx : Mucilaginous secretion surrounding many prokaryotic cell walls.

Glycolysis : Cycle in which glucose is broken down to form pyruvic acid.

Glycoprotein : Proteins featuring attachment of sugars, which are less than ten sugars long.

Glycosidase : The enzyme responsible for hydrolising a glucosidic linkage between two sugar molecules.

Golgi Body : Organelle comprising layers of flattened sacs, which absorbs and processes synthetic and secretory products from the endoplasmic reticulum and then secretes them to the cell's exterior or releases them into different parts of the cell.

Gonimoblast : Diploid filaments of carposporophyte (formed in Rhodophyta after fertilisation) that bear the carpospores.

Gorget : A small patch on the throat of an organism which is distinguished by its colour, texture and thickness quality.

Gradate : Graded, or stepped, of a sorus with the youngest sporangia at the apex and the older ones lower down.

Graft : Unification of the scion (shoot) of one plant and stock (root) of another plant.

Grain : (1) The texture of wood, produced by the kinds of xylem cells present. (2) The fruit of a member of the grasses.

Gram Stain : A differential stain that divides bacteria into two groups, as Gram positive and Gram negative, depending on the ability of the organism to retain crystal violet when decolorised with an organic solvent like ethanol.

Granular : Of a surface, finely mealy, covered with small granules.

Granum : The chloroplasts in vascular plants exhibit the presence of a series of stacked thylakoids, called granum.

Grassland : A community dominated by grass species.

Gravitational Water : After rain, the water draining into the pores of the soil is called gravitational water.

Greenhouse Effect : A process by which temperatures are warmed due to the accumulation of gases in an atmosphere which prevent the escape of absorbed inferred radiation from the sun.

Gregarious : Growing in groups or colonies.

Ground Meristem : The mersitem producing all the primary tissues of the plant except the epidermis and the stele.

Growth : An increase in the number of cells, and the size and constituents present in the cells.

Growth Factor : Organic compound essential for growth which is required in trace amounts, and which cannot be synthesised by the organism itself.

Growth Rate : The rate at which growth occurs.

Growth Rate Constant : Slope of log10 of the number of cells per unit volume plotted against time.

Growth Yield Coefficient : Quantity of carbon formed per unit of substrate carbon consumed.

Guanine (G) : A nitrogenous base, one member of the base pair GC (guanine and cytosine) in DNA.

Guard Cells : Pair of cells which surround a stomate and regulate its size by altering their shape.

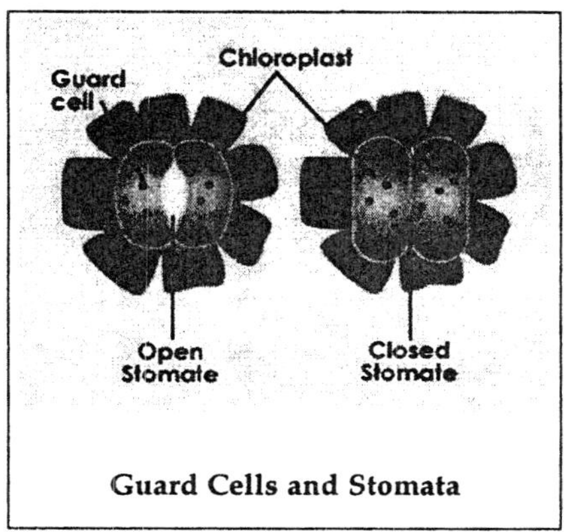

Guard Cells and Stomata

Gular Pouch : A bare sac or pouch that can be expanded to accommodate a large prey, or for the show off during courtship display.

Guttation : Exudation of water from the leaves in the form of droplets due to root pressure.

Gyandromorph : Organisms that have both male and female cells and therefore express both male and female characteristics.

Gymnosperm : A type of plant that bears seeds not enclosed by any type of specialised structure. Conifers are examples of gymnosperms.

Gynobase : An elongation or enlargement of the receptacle that supports the carpels or nutlets, as in many species of the Boraginaceae.

Gynobasic : Of a style, arising near the base of the gynoecium between the lobes of the ovary.

Gynoecium : The carpel (if solitary) or carpels of a flower collectively; the female part of the flower.

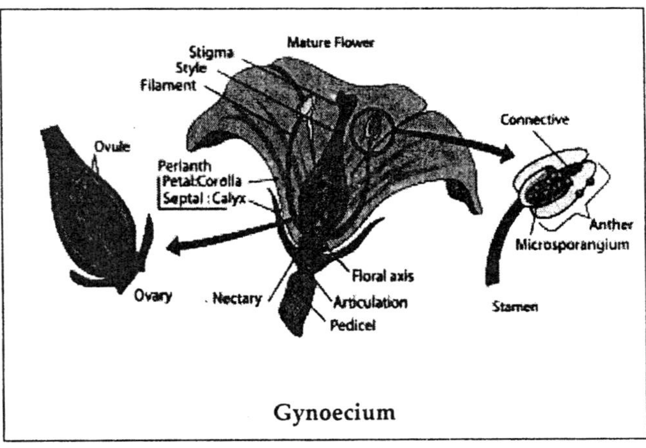

Gynoecium

Gynostemium : The central reproductive stalk of an orchid, which consists of a stamen and pistil fused together.

H

Habit : The general growth pattern of a plant. A plant's habit may be described as creeping, trees, shrubs, vines, etc.

Habitat : The natural environment in which the plant completes its life cycle.

Habitat Compression : When local population is forced or restricted within a set boundary, to accommodate more species.

Habitat Expansion : Increase in the habitat (overall area) distribution of the species.

Habitat Patch : A location that encompasses a distinct habitat type.

Habitat Selection : Habitats chosen over other habitats to suit climatic and environmental conditions.

Hair : An epidermal appendage, either unicellular or consisting of a single row or cells.

Half-inferior : Of an ovary, partly below and partly above the level of attachment of the perianth and stamens.

Halophile : An organism that thrives, or at least which can survive in a saline environment.

Halophyte : A plant adapted to living in a highly saline area; a plant that accumulates a high concentration of salt in its tissues.

Halotolerant : An organism that can survive in a saline environment, but does not require a saline environment for growth.

Haplobiontic **:** Having single, free-living phase in its life history.

Haploid **:** A single set of chromosomes (half the full set of genetic material) present in the egg and sperm cells of animals and in the egg and pollen cells of plants. Human beings have 23 chromosomes in their reproductive cells.

Haplotype **:** A specimen where doubt exists if the author actually handled the specimen mention in the description.

Hapten **:** A substance not inducing antibody formation, but which is able to combine with a specific antibody.

Haptera **:** Multicellular structures that form holdfasts of kelps.

Haptonema **:** Cellular extension with distinctive structure characteristic of some Prymnesiophyceae.

Hardwood **:** The wood of both dicot trees and shrubs are termed as hardwood. A dicot wood generally contains fibres.

Hastate **:** Spear-shaped; of a leaf, with a narrow, pointed lamina with two basal lobes spreading more or less at right angles to the petiole.

Hatchling **:** A young one that has just been hatched from an egg.

Haustaurium **:** Absorbing organ through which a parasitic plant absorbs nourishment from its host (pl. haustoria).

Haustorium **:** The absorbing organ of some parasitic or hemiparasitic plants through which substances pass from the host to the parasite.

Head (capitulum) **:** A dense cluster of more or less sessile flowers, e.g. in Asteraceae a group of florets sessile on a common receptacle.

Heartwood **:** Darker coloured non-living wood, whose cells have stopped conducting water.

Heath **:** (1) a plant community dominated by small, closely spaced shrubs, most of which have stiff and often small leaves; (2) a plant with small hard leaves, as in many Epacridaceae.

Helicoid **:** Coiled spirally like a spring or a snail shell.

Heliotropic **:** The movement of plant parts in response to a light source.

Hemi : A prefix meaning 'half'.

Hemiparasite : A parasitic (viz.) plant capable of limited photosynthetic production of sugars etc.

Hemizygous : Having only one copy of a particular gene. For example, in humans, males are hemizygous for genes found on the Y chromosome.

Herb : Generally any plant which does not produce wood, and is therefore not as large as a tree or shrub, is considered to be an herb.

Herbaceous : Herb-like, not woody; often applied to bracts, bracteoles or floral parts that are green and soft in texture.

Herbarium : Collection of plant specimens, which are pressed, dried, mounted on paper, identified and then labelled.

Hereditary Cancer : Cancer that occurs due to the inheritance of an altered gene within a family.

Hermaphrodite : Having both male and female reproductive organs, capable of forming male and female gametes on the same individual.

Hermaphroditic : Organism that have, as well as are capable of reproducing using both male and female reproductive organs.

Hetero : Prefix meaning 'dissimilar'.

Heterocyst : Specialised, thick-walled cell of some Cyanobacteria which may be site of nitrogen fixation.

Heterofermentation : Any fermentation where there is more than one main end product.

Heterogamous : Sexual reproduction where gametes are not identical in size or shape.

Heterokaryon : Hypha that contains at least two genetically dissimilar nuclei.

Heterokaryosis : Condition pertaining to certain cells in fungi, which feature two or more nuclei of different mating types.

Heterolactic Fermentation : A kind of lactic acid fermentation, wherein various sugars are fermented into different products.

Heteromorphic : Having life history phases morphologically different, having distinctly different gametophyte and sporophyte stages.

Heteromorphic, Heteromorphous : Existing in two or more easily recognisable forms.

Heterosporangiate : Producing two different kinds of sporangia, specifically microsporangia and megasporangia.

Heterosporous : Producing two different sizes or kinds of spores. These may come from the same or different sporangia, and may produce similar or different gametophytes.

Heterospory : Formation of both megaspores and microspores.

Heterostylous : Flowers with styles of different lengths, sizes or shapes in the same species.

Heterothallic : Hyphae that are incompatible with each other, thus requiring another compatible hypha to mate with, to form a dikaryon or a diploid.

Heterotrichy : Filamentous thallus having erect and prostrate parts.

Heterotrophic : Organisms that depend on other organisms for nutrition, as they are incapable of synthesising their own food.

Heterotrophic Nitrification : The oxidation of ammonium to nitrite and nitrate by heterotrophic organisms.

Heterozygosity : The presence of different alleles at one or more loci on homologous chromosomes.

Heterozygous : Possessing two different alleles of a trait on homologous chromosomes, which are situated at the same locus.

Hexa : Prefix meaning 'six'.

Hexose Monophosphate Pathway : A metabolic pathway involving the oxidative decarboxylation of glucose: 6 : phosphate.

Hibernal : Flowering or appearing in the winter.

Hibernation : To withdraw in a state of seclusion in a dormant condition.

Highly Conserved Sequence : DNA sequence that is very similar across several different types of organisms.

High-throughput Sequencing : A fast method of determining the order of bases in DNA.

Hilum : The scar on the seed coat where the seed was attached to the funicle.

Hip : A fleshy, berrylike fruit, as in some members of the *Rosaceae*.

Hirsute : Bearing coarse, moderately stiff, longish hairs.

Hirtellous : Pubescent with very small, coarse, stiff hairs.

Hispid : With stiff or rigid spreading hairs or bristles (dimin, hispidulous).

Histology : The study of tissues of organisms. It includes its structure, arrangement, functions, make up, etc.

Histones : Basic nuclear proteins forming complexes with DNA to form nucleosomes and then complexing further to form chromosomes.

Hoary : Covered with hairs so fine as to not be readily visible to the naked eye, giving the surface a pale greyish hue.

Hold Fast : Filament like organ of attachment present in algae that holds the algae to the substrate.

Holomictic : These are those lakes, wherein the water in them at some point of time will have a uniform temperature and density from top to bottom, thus allowing the lake waters to mix completely.

Holomorph : A fungus which consists of all sexual and asexual stages in its life cycle.

Holophyte : A plant maintained entirely by its own organs.

Holoplankton : Organisms that spend their entire existence as free-floating drifters.

Holosericeous : Covered with fine, silky hairs.

Holotype : A single specimen used as standard type to name, describe and illustrate, and represent a set of species and subspecies.

Homeobox : A short stretch of nucleotides whose base sequence is virtually identical in all the genes that contain it. Homeoboxes have been found in many organisms from fruit flies to human beings. In the fruit fly, a homeobox appears to determine when particular groups of genes are expressed during development.

Homeostasis : The process of maintaining internal stability of the physiological system of animals, in course of varying external conditions.

Homeothermy : The capacity to maintain the condition of being warm-blooded under all climatic situations.

Homo : Prefix meaning 'even' or similar'.

Homofermentation : A type of fermentation where there is only one type of end product generated.

Homokaryon : A fungal hypha containing nuclei which are genetically identical.

Homokaryosis : Condition in fungi, wherein all nuclei in the mycelium are genetically identical.

Homolactic Fermentation : A type of lactic acid fermentation, in which all sugars involved are converted into lactic acid.

Homolog : A member of a chromosome pair in diploid organisms or a gene that has the same origin and functions in two or more species.

Homologous Chromosomes : Diploid nucleus comprising a pair of chromosomes, one inherited maternally and the other paternally.

Homologous Recombination : Swapping of DNA fragments between paired chromosomes.

Homology : Similarity in DNA or protein sequences between individuals of the same species or among different species.

Homomorphic : All of the same kind or form.

Homonym : The same name; a combination (viz.) where the genus-species pair is exactly the same as an already existing combination, based on a different type (viz.).

Homosporous : Producing only one kind of spore in the sexual reproductive cycle, and hence each with gametophyte producing both male and female gametes.

Homostylous : Flowers with styles of similar length, size and shape in the same species.

Homothallic : Hyphae that are self : compatible, that is, sexual reproduction occurs in the same organism by meiosis and genetic recombination. Fusion of these hyphae lead to the formation of dikaryon or diploid.

Homozygote : An organism that has two identical alleles of a gene.

Homozygous : Possessing two identical alleles on a homologous chromosome pair at the same locus.

Hocked : Abruptly curved at the tip.

Hormone : Organic substances produced mostly in small amounts in one part of the organism and then transported to different parts of the organism, where it controls the growth and development of the organism.

Host : An organism on which a parasite lives and by which it is nourished; also applied to a plant supporting an epiphyte.

Humic Acid : Dark coloured organic material extracted from the soil by the use of reagents and which is precipitated by acid.

Humic Substances : High molecular weight substances formed by secondary synthesis reactions, for example, humic acid and fulvic acid.

Humification : The process of conversion of organic residues into humic substances by biochemical processes.

Humifuse : Spreading along or over the ground.

Humistrate : Lying on the ground.

Humus : Fine organic substance, composed of partial or full decomposed animal or plant matter, and found in soil.

Hyaline : Thin, translucent or transparent.

Hybrid : The offspring of genetically different parents (in a flora, usually applied where the parents are of different species).

Hybrid Sterility : Post-zygotic isolation process, wherein a hybrid zygote develops into an adult, however, is incapable of forming fertile gametes.

Hybrid Swarm : A variable population resulting from crossing and segregation amongst the offspring of a hybrid or hybrids, including one or both of the parent taxa.

Hybridisation : Natural or artificial construction of a duplex nucleic acid molecule by complementary base pairing between two nucleic acid strands derived from different sources.

Hydathode : Specialised leaf structure located at the leaf's tip, from which the water is forced out when root pressure increases.

Hydrocarbon : An organic compound containing carbon and hydrogen only.

Hydrogen Oxidising Bacterium : These are bacteria that oxidise hydrogen for energy and synthesise carbohydrates, using carbon dioxide as their source of carbon in the absence of other organic compounds.

Hydrophyte : A plant growing submerged in water, sometimes partly emergent.

Hydrophytic : Adapted to growing in water.

Hygroscopic : Capable of expanding or contracting in response to presence or absence of water or atmospheric moisture.

Hygroscopic Water : Water chemically adhering to soil particles due to which they are unavailable to plants.

Hymenium : Layer of fertile cells producing spores in a fungus fruiting body.

Hyoid Apparatus : A veterinary anatomy term for the upper throat bones of the tongue and connective tissues.

Hypanthium : A cuplike or tubular structure formed above the base, and often above the top, of the ovary with the stamens and perianth parts inserted on the rim. e.g. as in Onagraceae and some Myrtaceae.

Hyperparasite : Parasite that feeds on another parasite.

Hyperthermophile : An organism that thrives in temperatures ranging around 80 degrees Celsius or more.

Hypha : n.Threadlike filaments that form the mycelium (body) of a fungus.

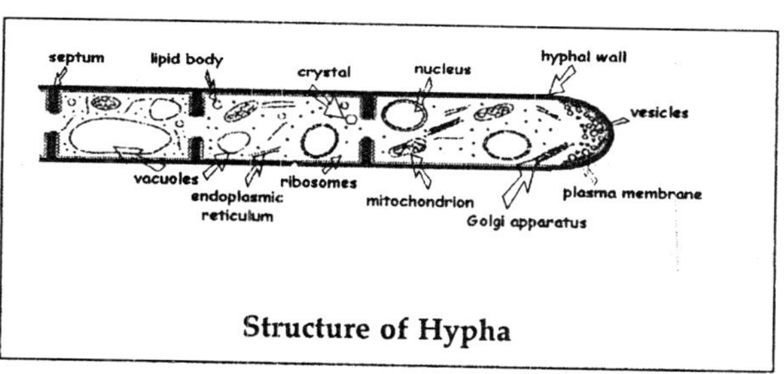

Structure of Hypha

Hypo : Prefix meaning 'beneath' or 'under'.

Hypocone : Portion of dinoflagellate cell posterior to the cingulum.

Hypocotyl : Portion between the cotyledon and the radicle in a seedling or embryo.

Hypodermis : Cell layer following the epidermal layer and distinct from the cortical parenchyma cells in some plants.

Hypogynous : Condition featuring attachment of flower parts below the ovary, e.g. of sepals, petals and stamens.

Hypogynous Cell : Cell directly beneath the carpogonium in the Rhodophyceae.

Hypolimnion : This is the dense, bottom layer of water, that lies below the thermocline, in a thermally stratified lake.

Hypolithic : Living on lower surface of rocks.

Hypotheca : Part of diatom frustule consisting of hypovalve and hypocingulum.

Hypovalve : Flattened or covex plate of diatom frustule opposite the epivalve.

I

Identical Twin **:** Twins produced by the division of a single zygote; both have identical genotypes.

Idioblast **:** A specialised cell which differs from its neighbours in size, structure and function.

Illuviation **:** Repositioning of soil removed from one horizon to another.

Imago **:** Sexually mature adult stage in the life of certain insects.

Imbricate **:** Overlapping like the shingles on a roof; either vertically or spirally where the lower piece covers the base of the next higher, or laterally where at least one piece must be wholly external and one internal.

Immobile Essential Element **:** Element that cannot be removed from mature tissues. This means if young tissues become deficient in these elements they develop a deficiency, even though this element is present in the older tissues.

Immobilisation **:** Conversion of an element from inorganic to organic form.

Immunity **:** The protection mechanism against infections caused by microorganisms or toxins, that is inherent in the body.

Immunoblot **:** The technique for analysing or identifying proteins via antigen: antibody specific reactions.

Immunofluoresence **:** The technique to determine the location of an antigen or antibody in a tissue section or smear by fluorescence.

Immunogen **:** A substance that has the capacity to bring about an immune response.

Immunoglobulin : A protein which has antibody activity.

Immunotherapy : Using the immune system to treat disease, for example, in the development of vaccines. May also refer to the therapy of diseases caused by the immune system.

Imparipinnate : Term describing a pinnate leaf with a single terminal leaflet, and therefore usually with an odd number of leaflets.

Imperfect : Describes a flower that has stamens or pistils but not both.

Imperfect Flower : Flowers lacking either carpels or stamens or both.

Imperfect Fungi : Those fungi that do not sexually reproduce or their sexual reproduction behaviour has never been monitored.

Impermeable Membrane : Membranes that do not permit the passage of any substances across them.

Implicate : Twisted together, intertwined.

Impressed : Sunk or immersed below the level of the surface.

Imprinting : A phenomenon in which the disease phenotype depends on which parent passed on the disease gene. For instance, both Prader-Willi and Angelman syndromes are inherited when the same part of chromosome 15 is missing. When the father's complement of 15 is missing, the child has Prader-Willi, but when the mother's complement of 15 is missing, the child has Angelman syndrome.

In : Prefix meaning 'not' or 'inwards'.

In Situ Hybridisation : Use of a DNA or RNA probe to detect the presence of the complementary DNA sequence in cloned bacterial or cultured eukaryotic cells.

In Vitro : Carrying out growth of cells in artificially maintained media, such as test tubes, flasks, etc. instead of inside a living organism.

Inbreeding : Process of individuals with common ancestry mating together.

Inbreeding Depression : Condition in which individuals with common ancestry exhibit low fertility and poor performance.

Incipient Plasmolysis : The point at which the protoplasm just begins to stop exerting pressure on the cell wall, when the plant cell membrane shrinks after losing water.

Incised : Cut, often deeply, usually irregularly, but seldom as much as one-half the distance to the midrib or base.

Included : Enclosed, not protruding, e.g. of stamens not projecting beyond the perianth or of valves which do not extend beyond the rim of a capsular fruit.

Incomplete Flower : A flower lacking carpels, sepals, petals, and/or stamens.

Incumbent : A term referring to seeds in which the embronic root is wrapped around and lies adjacent to the back of one of the two cotylodons.

Incurved : Of leaf margins, curved inwards or upwards.

Indefinite : Variable in number; numerous; of an inflorescence, not terminating in a flower.

Indehiscent : Not opening by itself, said of a seed pod.

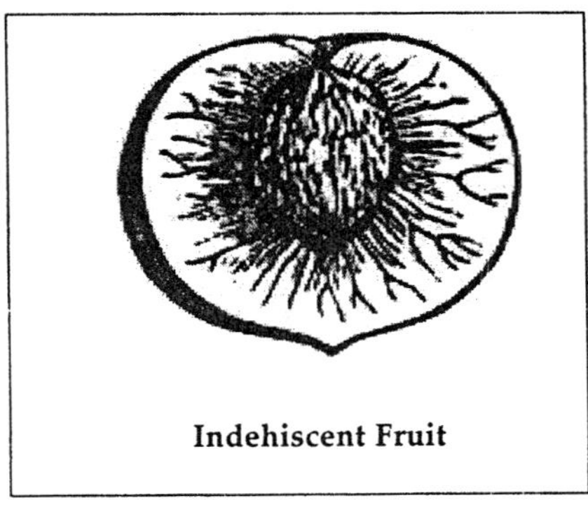

Indehiscent Fruit

Independent Assortment : During meiosis each of the two copies of a gene is distributed to the germ cells independently of the distribution of other genes.

Indeterminate : (1) term describing growth or branching in which the terminal bud persists and produces successive lateral branches; (2) of an inflorescence of part of an inflorescence (=blastotelic), not ending in a flower, i.e. ending in a non-floral bud, e.g. a thyrse, raceme or spike.

Indeterminate Growth : A type of growth in which a plant (or other organism) continues to grow throughout its life, until death.

Indigenous : Native, occurring naturally in an area; hence indigene.

Inducible Enzyme : An enzyme generated in response to an external factor.

Indumentum : Any covering of a plant surface, especially hairs and scales.

Induplicate : The margins bent inwards, and the external faces of these edges applied to each other, without twisting.

Indurated : Hardened and toughened.

Indusium : The protective membrane covering a sorus, not originating from the margin of the lamina.

Infection : Invasion and multiplication of microorganisms in body tissues, leading to various diseases and disorders.

Infection Thread : The tube in root hair, through which rhizobia reach and infect roots.

Inferior : Situated below another organ or part.

Inferior Ovary : An ovary below the level of attachment of the perianth parts and stamens and completely fused with the hypanthium or at most with a free summit; if less fused. The floral parts of a flower with such an ovary are said to be epigynous.

Inflexed : Turned abruptly or bent inwards.

Inflorescence : A general term for the flower-bearing system of a plant, and more particularly for portions of such systems separated from one another by vegetative portions of the plant.

Informed Consent : An individual willingly agrees to participate in an activity after first being advised of the risks and benefits.

Infra : A prefix meaning below or beneath.

Infrared (IR) : The portion of the electromagnetic spectrum whose wavelength ranges from 0.75 microns to 1 millimeter.

Infraspecific : Of taxonomic divisions of a lower rank than species; similarly infrageneric, infrafamilial etc.

Infructescence : The inflorescence in the fruiting stage; the arrangement of fruits, including peduncle, pedicels, bracts and fruit.

Inherit : In genetics, to receive genetic material from parents through biological processes.

Innovation : A new vigorous shoot, carrying on the continued growth of the plant.

Inoculate : To treat a medium with microorganisms for the purpose of creating a favourable response.

Inoculum : The material used to introduce an organism into a certain medium for growth.

Insectivore : An organism that feeds chiefly on insects.

Insectivorous : Trapping and feeding on insects and, by extension, other small invertebrates.

Inserted : Attached to or growing upon; hence insertion, the place or mode of attachment.

Insertion : A chromosome abnormality in which a piece of DNA is incorporated into a gene and thereby disrupts the gene's normal function.

Insertion Sequence : The simplest possible type of transposable elements.

Insipid : Flat, rather tasteless; not tart.

Integration : The process by which a DNA molecule becomes incorporated into another genome.

Integument : Outermost wall of the ovule, which develops into the seed coat. In angiosperms, the ovule has two integuments, while in gymnosperms, a single integument is seen.

Intellectual Property Rights : Patents, copyrights, and trademarks.

Inter : A prefix meaning between or among.

Intercalary : Of a meristem (growing region), situated between regions of permanent tissue, e.g. at the base of nodes and leaves in many monocotyledons.

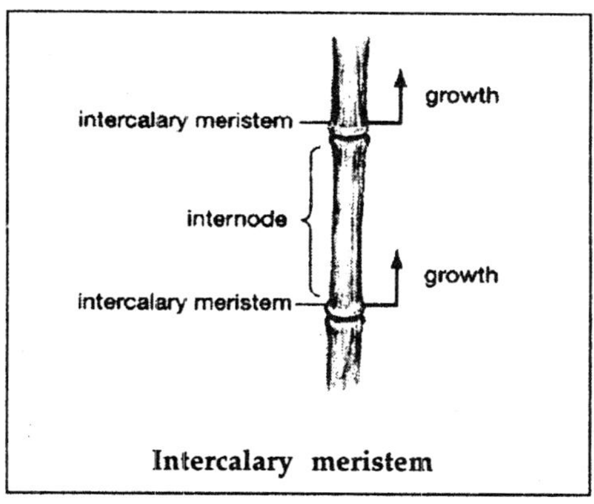

Intercalary meristem

Intercellular Space : Space present between two adjacent cells.

Interference : One crossover event inhibits the chances of another crossover event. Also known as positive interference. Negative interference increases the chance of a second crossover.

Intergenicula : Non–flexible, calcified segments between flexible joints in articulated corallines and *Halimeda*.

Interjugary : Of glands, present on the rachises of bipinnate leaves between the junction of pairs of pinnae or of pinnules. as in some Acacia species.

Intermediate Leaves : Leaves that develop after the juvenile and before the mature leaves in plants which have dimorphic or trimorphic foliage.

Intermediate-day Plant : Plants characterised by two critical photo periods. This means the plant will not flower during too short or too long days.

Internode : The region of a stem between two nodes, when there is no branching of the vascular tissue.

Interpetiolar : Of stipules, between the petiole bases of two opposite leaves.

Interphase : Phase of cell cycle which is not cell division but encompasses phases such as G1, S, G2. Here, the cell prepares for cell division.

Interrupted : Not continuous, with gaps.

Interrupted Inflorescence : One with flowers in distinct clusters and with bare axis or stem between the clusters.

Interspecies Hydrogen Transfer : The process of hydrogen production and consumption reactions, occurring by the interaction of various microorganisms.

Interstitial Skin : The skin found between the scales of a snake.

Intertidal : Occurring between high and low tide levels, exposed at low tide.

Intramarginal : Situated inside the margin but close to it, e.g. of veins in the leaves of many Myrtaceae.

Intricate : Of plants, with many entangled branches.

Intrinsic Protein : Protein deeply integrated into the membrane, which cannot be discarded from the membrane easily.

Introduced : Not indigenous (viz.), not native to the area in which it now occurs, brought in by accident or design.

Introduced Species : Organism that would not normally occur but have been introduced in the habitat.

Introgression : The modification of species by the back-crossing of the hybrids with the parent species.

Intron : DNA sequence that interrupts the protein-coding sequence of a gene; an intron is transcribed into RNA but is cut out of the message before it is translated into protein.

Introrse : Turned or opening inward toward the axis as an anther toward the centre of a flower.

Invagination : The process of forming a pocket by turning in on itself, as in the floral axis of figs (e.g. Ficus species where the minute flowers and fruits are actually inside the swollen inflorescence stem, the 'fig'); the resulting multiple fruit is a syconium.

Involucel : A secondary involucre as in the *Apiaceae*.

Involucre : (1) a whorl or several whorls of bracts surrounding a flower or an inflorescence (as around the head in many Asteraceae); (2) a layer of tissue enveloping a particular structure, such as sporangia in many ferns, e.g. Hymenophyllaceae.

Involute : With both edges inrolled toward the midnerve on the upper surface.

Irregular : Describes a flower that is not radially symmetric, the similar parts of which are unequal in size or form.

Irrigation : A system designed to deliver water to plants.

Iso : Prefix meaning 'same' or 'equal'

Isobilateral (isolateral) : Having structurally similar upper and lower surfaces.

Iso-enzyme : When two different enzymes, which maybe different in their composition, act as catalysts for the same reaction, or set of reactions.

Isogamous : Having gametes which are morphologically alike.

Isogamy : Sexual reproduction taking place between gametes that are similar in size. Seen in certain fungi and algae.

Isolating Mechanism : Prevention of breeding between species due to behaviour, morphology, genetics or a geographical barrier.

Isolation : A procedure wherein a pure culture of an organism is obtained from a sample or an environment.

Isomorphic : Morphologically similar, having similar looking gametophyte and sporophyte phases of life history.

Isomorphous Substitution : The substitution of an atom by a similarly sized atom of lower valence, in a crystalline clay sheet.

Isotype : A duplicate of the holotype, a specimen made from the same collection as the holotype.

J

Jaccard's Coefficient : An association coefficient of numerical taxonomy, which is the proportion of characters that match, excluding those that both organisms lack.

Joint : (1) an articulation, as in a 1-foliolate leaf, (2) a segment of some cladodes, as in many Cactaceae; (3) a node.

Jointed Stems : Stems made up of one jointed, three dimensional and ribbed parts, more like a clump from where branches shoot.

Jugary : Of glands, present on the rachises of bipinnate leaves at the junctions of pairs of pinnae or of pinnules, as in some Acacia species.

Jugate : Yoked together in a pair, mostly of pairs, of pinnae in a pinnate leaf; also as a suffix, bijugate, 4-jugate etc.

Junciform : Rushlike in appearance.

Jungle : A dense growth of various plants where many organism can thrive. Forest is another term used for a jungle.

Junipers : This term refers to members of the Family Cupressaceae, and are characterised needle-like leaves in juveniles and scale-like leaves and cones in the adults.

Junk DNA : Stretches of DNA that do not code for genes; most of the genome consists of so-called junk DNA which may have regulatory and other functions. Also called non-coding DNA.

Juvenile Leaves : The first-formed leaves, especially when they differ from the mature leaves.

K

K- Strategy : Ecological strategy where organisms depend on adapting physiologically to the resources available in their immediate environment.

Karyogamy : Fusion of two gametes of the nuclei after plasmogamy (protoplasmic fusion).

Karyokinesis : Process of division involving series of active changes in the nucleus of a cell.

Karyotype : A photomicrograph of an individual's chromosomes arranged in a standard format showing the number, size, and shape of each chromosome type; used in low-resolution physical mapping to correlate gross chromosomal abnormalities with the characteristics of specific diseases.

Keel : (1) a ridge like the keel of a boat, usually on the back of an organ; (2) the two fused anterior petals of the pea flower.

Kelp : Member of brown algal order Laminariales.

Kernel : A grain or seed as of corn, wheat, etc.; the inner softer part of a nut or fruit stone.

Key : Tools used to identify unfamiliar plants. Comprise mainly of pairs of choices.

Kilobase (kb) : Unit of length for DNA fragments equal to 1000 nucleotides.

Kinetochore : During late prophase, some specialised protein complexes are developed on the vertical faces of a centromere, and are called kinetochore.

Kingdom : Highest level of classification category.

Kino : Reddish exudate from the bark or wood of some trees.

Kleptoparasitism : A parasitic characteristic of opportunistically stealing food and/or nests from other organisms.

Kleptotype : A fragment removed from or stolen from the type; a highly illegal and immoral procedure.

Knockout : Deactivation of specific genes; used in laboratory organisms to study gene function.

Knot : Projection of plant tissue in the stem, root, etc. especially when swollen.

Koch's Postulates : Laws given by Robert Koch which prove that an organism is the causative agent of a disease.

Kranz Anatomy : A specialised anatomy associated with the C4 carboxylation pathway in plants. The vascular system is associated with at least two distinct photosynthetic cell types usually arranged in concentric layers. The cell types differ in ultrastructure and in function. Found in some grasses (Poaceae) and chenopods (Chenopodiaceae).

L

Labellum : The distinctive median petal in Orchidaceae, usually differing in size and shape from those either side.

Labial : Labial refers to the lips.

Labium : Lip; hence labiate (pl. labia).

Lacerate : As if torn; irregularly cut or cleft.

Laciniate : Deeply, usually irregularly divided into very narrow, pointed segments.

Lacuna : A gap, a space enclosed by but free from veins; hence lacunose, lacunate.

Lag Phase : The time period when there is no increase in the number of microorganisms, seen after inoculation of fresh growth medium.

Lamella : Seen in plants as the layers of protoplasmic membranes in chloroplast that contain photosynthetic pigments.

Lamina : A thin, flat organ or part, especially the expanded blade of a leaf; hence laminate (dimin. lamella, lamellate).

Laminaran : Storage product of Phaeophyta, branched polymer of polysaccharide made up of glucans, soluble in the cell.

Laminate : Broadened into a lamina.

Lanate : Clad in woolly, usually intertwined, hairs.

Lanceolate : Lance-shaped, of a plane several times longer than wide, widest in the basal third, tapering gradually towards the tip, more rapidly towards the base.

Lanulose : With very short hairs, minutely downy or woolly.

Last Common Ancestor : This term refers to the most recent known and shared common ancestor between two species, as well as individuals.

Late Wood : Also referred to as summer wood, this is the wood formed late in season, in the secondary xylem. It usually comprises of narrow tracheids in gymnosperms and few or no vessels in angiosperms.

Lateral : Refers to the side location or view.

Lateral Bud : A bud located along the side of a stem, usually in an axillary position, by which the stem branches.

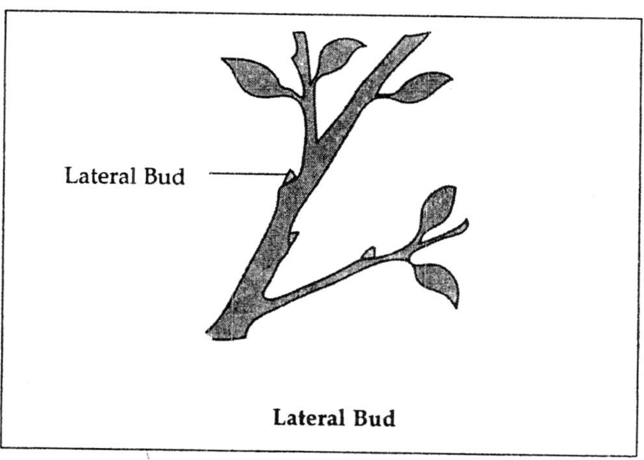

Lateral Bud

Lateral Meristem : Meristem located on the lateral portion of a plant, a point at which secondary growth may occur.

Lateral Roots : The scores of tiny roots stemming from the tap root.

Latex : A fluid exuded from cut surfaces of the leaves and stems of some plants, usually milky, sometimes yellowish and watery, e.g. as in many Moraceae and Apocynaceae.

Laticifer : Specialised ducts or cells that bear resemblance to vessels. These form a network of cells in the phloem and other plant parts that secrete latex.

Latrorse **:** Turned sideways; of anthers, dehiscing longitudinally on the side.

Lax **:** Loosely arranged or distantly placed.

Layering **:** Propagation method whereby roots are encouraged to form from branches of a plant. After roots are established, the branch is cut off and planted by itself.

Leaching **:** Removal of metals from ores by the help of microorganisms.

Leaf **:** An organ found in most vascular plants; it consists of a flat lamina (blade) and a petiole (stalk). Many flowering plants have additionally a pair of small stipules near the base of the petiole.

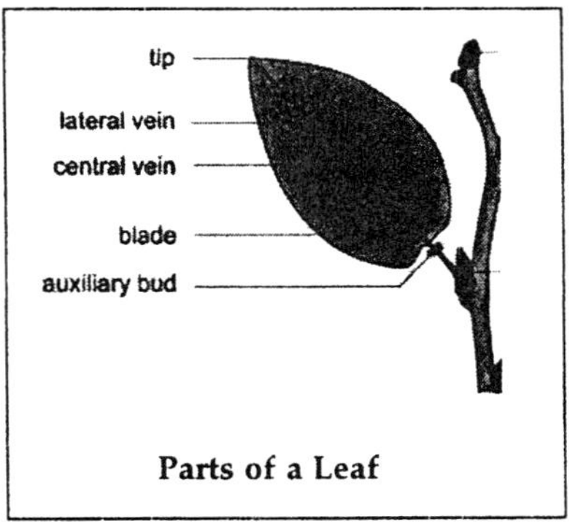

Parts of a Leaf

Leaf Mold **:** Partially decomposed leaves that are used in soils for the addition of organic nutrients.

Leaf Scar **:** Portion of the stem, wherein the leaf was attached, before its abscission.

Leaf Sear **:** The scar left on the stem by the petiole of the leaf after the leaf falls.

Leaf Trace **:** The strand of vascular tissue which connects the leaf veins to the central vascular system of the stem.

Leaf-gap **:** The break in the vasculature of the stem where the leaf-trace(s) leaves the stem to enter the petiole.

Leaflet : In a compound leaf, the individual blades are called leaflets.

Leaf-opposed : Arising from the stem opposite a leaf-base, as do some tendrils or inflorescences.

Lectins : Plant proteins with a high affinity for specific sugar residues.

Lectotype : A specimen or similar element selected from the original material to serve as the nomenclatural type when a holotype was not originally designated, or as long as it is missing.

Leghemoglobin : Red coloured pigments rich in iron, which are produced in root nodules during symbiotic association between rhizobia and leguminous plants.

Legume : (1) a pod, a dry dehiscent fruit formed from one carpel and having two longitudinal lines of dehiscence, (2) a member of the family Fabaceae.

Legume Fruit

Lemma : In grasses, the lower and usually larger of the two bracts of the floret.

Lenticel : Spongy cluster of cells located in the bark of woody plants, which allow gas exchange between the external atmosphere and interior of a plant.

Lenticular : A 3-dimensional shape like a doubly convex lens, circular in outline.

Lepidote : Covered with small scurfy scales.

Lepto : Prefix meaning 'thin', 'slender' or 'narrow'.

Leptosporangiate : Of ferns, having sporangia with walls consisting of a single layer of cells; the sporangium originating from a single cell.

Leucoplast : Colourless, usually starch-containing plastid.

Liana : A herbaceous or woody, usually perennial, climbing vine that roots in the ground and is characteristic especially of tropical forests.

Library : An unordered collection of clones (i.e., cloned DNA from a particular organism) whose relationship to each other can be established by physical mapping.

Lichen : Fungi living in symbiotic union with algae. They appear either leaflike, crustlike or in the form of branching trees, rocks, etc.

Life Cycle : Phases of life that an organism go through starting from birth, to sexual maturity, till death.

Ligand : A molecule, ion or group of molecules or ions, bound to the central atom by means of a chelate or coordination compound.

Light Compensation Point : The point where the rate of respiration is higher than the rate of photosynthesis, which usually occurs at about 1% of sunlight intensity.

Light Reactions : The first of two main steps in photosynthesis where solar energy is converted into chemical energy.

Light-dependent Reactions : Chain of chemical reactions involving the conversion of light energy into chemical energy with the assistance of chlorophyll pigment.

Light-independent Reactions : Cyclic sequence of chemical reactions utilising carbon dioxide and energy released during the light-dependent reactions. These reactions are independent of light, and take place in the stroma of chloroplasts.

Lignin : Hard material is cellulose plant cell walls used for support in terrestrial plants.

Lignotuber : A woody swelling, partly or wholly underground, at the base of certain plants and containing numerous cortical buds, as in many eucalypts.

Ligulate : (1) Describing a floral head in the *Asteraceae* that contains only ray flowers, or ligules; (2) strap-shaped.

Ligule : (1) a variously shaped appendage facing towards the base of a leaf (especially in grasses), petiole, or perianth segment; (2) the strap.

Limb : A part towards the edge as opposed to the central part or disc; the expanded portion of a sepal, petal or leaf.

Lime : Soil amendment containing high levels of calcium compounds, like calcium carbonate and other such mineral which are used to neutralise soil acidity, and provide calcium for plant growth.

Linear : Long and narrow with more or less parallel sides, more than 12 times as long as broad.

Linear-lanceolate : More than 12 times as long as broad and broadest in the lowest third and tapering to the apex.

Linear-oblanceolate : More than 12 times as long as broad and broadest in the upper third and tapering to the base.

Lineate : Marked with parallel lines.

Linkage : The proximity of two or more markers (e.g., genes, RFLP markers) on a chromosome; the closer the markers, the lower the probability that they will be separated during DNA repair or replication processes (binary fission in prokaryotes, mitosis or meiosis in eukaryotes), and hence the greater the probability that they will be inherited together.

Linkage Disequilibrium : Where alleles occur together more often than can be accounted for by chance. Indicates that the two alleles are physically close on the DNA strand.

Linkage Map : A map of the relative positions of genetic loci on a chromosome, determined on the basis of how often the loci are inherited together.

Linked Genes : Genes situated close together on the same chromosome that crosses over only rarely.

Lipid : Hydrophobic and water insoluble compounds, such as waxes, fats, oils, etc.

Lipopolysaccharide (LPS) : Complex lipid structure containing sugars and fatty acids, which is commonly found in most Gram negative bacteria.

Lithophyte : A plant growing on rocks.

Lithotroph : An organism that uses inorganic substrate such as ammonia or hydrogen to act as electron donors in energy metabolism. They maybe chemolithotrophs or photolithotrophs.

Litter : The surface layer of forests which is laden with leaves, twigs, fruits etc.

Littoral : On or growing near the seashore.

Lobe : Any portion of a structure, especially if rounded, which is at least partially set off from the remainder.

Lobed : More or less deeply cut but not as far as the midrib.

Localise : Determination of the original position (locus) of a gene or other marker on a chromosome.

Locule : Hollow situated within a sporangium or ovary.

Locule, Loculus : A compartment or cavity of an organ; hence loculate.

Loculicidal : Said of a capsule, longitudinally dehiscent through the ovary wall at or near the centre of each chamber or locule.

Loculicidal Dehiscence : In capsules, dehiscence in median lines through the walls of the loculi rather than at the partitions between the loculi or at the placentas.

Loculus : A more or less closed cavity, containing the pollen in anthers and the ovules in an ovary. pl. loculi.

Locus : The position on a chromosome of a gene or other chromosome marker; also, the DNA at that position. The use of locus is sometimes restricted to mean expressed DNA regions.

Lodicule(s) : One or two scale-like structures below the stamens and ovary of a grass and regarded as a reduced perianth.

Loment : A legume which is constricted between the seeds.

Lomentum : A legume that breaks transversely into usually 1-seeded indehiscent articles when mature.

Long Day Plant : A plant that flowers only when daily sunlight hours surpass some critical level.

Long Shoot : These shoots feature tiny papery leaves; as observed in conifers.

Long-distance Transport : Transportation of substances from one cell to another cell, which is situated at a far away location.

Longitudinal : Of venation, with several veins extending from the base to the apex of the lamina but the veins not more or less parallel with each other.

Long-Range Restriction Mapping : Restriction enzymes are proteins that cut DNA at precise locations. Restriction maps depict the chromosomal positions of restriction-enzyme cutting sites. These are used as biochemical "signposts," or markers of specific areas along the chromosomes.

Lophotrichous : An organism that has a tuft of flagella that is polar in nature.

Lumen : Inner portion of cell structures such as vacuole, vesicle, resin duct or oil chamber.

Lunate : The shape of a half-moon or crescent.

Lurid : Pale brown to yellowish-brown.

Luxury Uptake : Uptake of nutrients in excess of what is required by an organism for its normal growth.

Lyrate : Lyre-shaped, of pinnatifid or pinnatisect leaves with the terminal lobes much larger than the basal ones.

Lysis : The rupture and destruction of a cell, resulting in loss of cellular contents.

Lysogeny : An association where a prokaryote contains a prophage and the virus genome is replicated in sync with the chromosome of the host.

Lysosome : A cell organelle which contains lytic enzymes.

M

Machaerantheroid : Having involucral bracts with recurved tips.

Macro : Prefix meaning large or long.

Macroalgae : Large, macroscopic algae, seaweeds.

Macronutrient : A substance required in large amounts for normal growth of an individual.

Macrophyllous : Having large leaves.

Macropore : Larger soil pores from which water drains readily by gravity.

Macrothallus : Large, conspicuous phase of organisms's life history.

Macula : Spot or blotch; hence maculate.

Magnetosome : Small particles of magnetite, which is a compound containing magnesium, present in cells that exhibit magnetotaxis.

Magnetotactic Bacteria : Bacteria that orient themselves according to the earth's magnetic field due to the presence of the magnetosomes.

Magnoliid : Any member of the basal assemblage of flowering plants.

Malathion : Synthetic pesticide used to control a wide range of biological pests.

Mallee : (1) a growth form in which many stems arise from a lignotuber, usually applied to eucalypts; (2) of a plant community dominated by mallee eucalypts.

Mamillate : With nipple-like projections.

Mangrove : A shrub or small tree growing in salt or brackish water and often with pneumatophores or aerial roots.

Manicate : With a thick, interwoven pubescence.

Mannoxylic : Wood in which there is a great deal of parenchyma tissue among the xylem is called mannoxylic. Cycads and pteridosperms have mannoxylic wood.

Manure : Animal excreta, with or without a bedding of litter at various stages of decomposition. It's normally considered to be a good fertiliser.

Marcescent : Withering but still persistent as with petals and sepals or the basal leaves of some plants.

Mare : Mare is a female horse that has attained the age of more than four years.

Marginal : (1) at or very close to the margin; (2) of placentation, with the placenta along the margin of a simple ovary, as in many legumes.

Maritime : Belonging to the sea; confined to the seacoast.

Marsh : A waterlogged area; swampy ground without trees.

Mass Flow (nutrient) : The movement of solutes in relation to the movement of water.

Mass Selection : Plant breeding procedure involving formation of an amalgamated population via selective harvesting of individuals from a population which is heterozygous.

Mass Spectrometry : An instrument used to identify chemicals in a substance by their mass and charge.

Massula : Rounded mass of hardened cytoplasmic foam containing one or more spores in the Salvineales.

Mastigoneme : Fine hair like appendage of a flagellum.

Maternal Inheritance : Condition in which offspring receives extranuclear material from the female gamete.

Matinal : Blooming in the early morning.

Mating Types : Biochemical attributes that differentiate one mating type from the other. Gametes associated with the same mating type cannot fuse, and require compatible mating types for syngamy.

Mealy : Describing a surface that is covered with minute, usually rounded particles.

Medial : Attached near or at the middle, especially midway between costa and margin.

Medium : A source where microorganisms are grown.

Medulla : Parenchymatous tissue within the vascular cylinder.

Medullated Protostele : A cylindrical stele with a nonvascular, parenchymatous centre, with the phloem around the outside only, also known as an ectphloic siphonostele.

Medusa : The sexual stage in the life cycle of a coelenterate, such as a jellyfish or a hydra, in which it is free-swimming.

Mega : Prefix meaning 'large'.

Megabase (Mb) : Unit of length for DNA fragments equal to 1 million nucleotides and roughly equal to 1 cM.

Megagametophyte (also Macrogametophyte) : The gametophyte developing vegetatively from the megaspore of a heterosporous plant, female.

Megaphyll : Present in all seed plants and ferns. It is a leaf that has evolved from a branch system and is characterised by branching veins.

Megasporangium (also Macrosporangium) : The sporangium in heterosporous plants where the megaspores develop (pl. megasporangia).

Megaspore(also Macrospore) : The spore in heterosporous plants that gives rise to a female gametophyte and is generally larger than the microspore; the spore usually not shed but remaining on the parent plant and developing in situ.

Megasporocyte : Diploid cells undergoing meiosis to form megaspores.

Megasporophyll (also Macrosporophyll) : Specialised leaf bearing or subtending one or more megasporangia.

Meiosis : It is the process of nuclear division in a cell, in which the total number of chromosomes is reduced to half. Meiosis results in the formation of gametes in animals and spores in other organisms. Before the process begins the DNA in the original cell are replicated during a phase called S-phase and this is similar to that in mitosis. Once the replication is completed, two cell divisions separate the replicated chromosomes into four haploid gametes or spores.

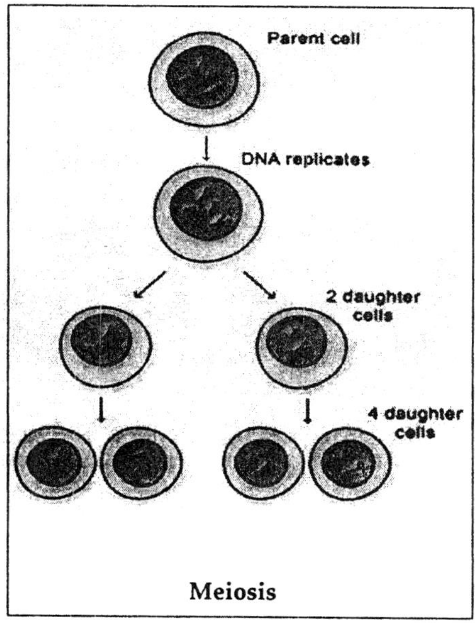

Meiosis

Membranaceous, Membranous : Having the texture of membrane; a thin, soft, pliable layer which is more or less translucent.

Membrane : A thin, soft, flexible, +/- translucent piece of tissue; hence membranous.

Membranous : Thin, flexible and more or less translucent, like a membrane.

Mendelian Inheritance : One method in which genetic traits are passed from parents to offspring. Named for Gregor Mendel, who first studied and recognised the existence of genes and this method of inheritance.

Mendelian Trait : Transmission of hereditary traits from parent organisms to their offsprings, the trait mainly emphasises on a single locus.

Meniscoid : Venation in which the veins are arranged pinnately from a main vein and anastomose with the neighbouring group to form a single excurent vein.

Mericarp : One segment of a fruit that breaks at maturity into units derived from the individual carpels, sometimes called a coccus.

Meristele : The individual vascular bundle of a complex stele such as a dictyostele.

Meristem : Group of undifferentiated cells from which new tissues are produced. Most plants have apical meristems which give rise to the primary tissues of plants, and some have secondary meristems which add wood or bark.

Meristoderm : Superficial, outer layer of dividing cells in brown algae in the order Laminariales.

Merophytes : Group of cells which have all been produced from the same initial cell. Leaves and stems in particular are often built from specific patterns of merophytes.

Meroplankton : Organisms that spend part of their life cycle as plankton and part on the benthos.

Merous : A suffix indicating that the number of parts in each floral whorl is divisible by the same basic number, e.g. a 5-merous flower is one with the number of sepals, petals and stamens divisible by 5; e.g. 5 sepals, 10 or 5 petals, and 5, 15 or 20 stamens. The number of carpels and their styles or stigmas often does not conform to the basic number.

Mesic : Describes a habitat that is generally moist throughout the growing season.

Meso : Prefix meaning middle.

Mesocarp : Central portion of the fruit wall, which is sandwiched between the outer exocarp and inner endocarp.

Mesokaryotic : Having chromosomes that persist in condensed form at all times, the condition of the dinoflagellate nucleus.

Mesomorphic : Soft and with little fibrous tissue, but not succulent.

Mesophile : An organism that thrives in temperatures ranging from 15 - 40 degrees Celsius.

Mesophyll : Tissues (parenchyma or chlorenchyma) situated between the epidermal layers of the leaf.

Mesophytic : Adapted to growing under medium or average conditions, especially relating to water supply.

Messenger RNA (mRNA) : Single stranded RNA molecule carrying genetic information from the DNA template to the site of protein synthesis.

Metaphase : A stage in mitosis or meiosis during which the chromosomes are aligned along the equatorial plane of the cell.

Methanogenesis : The production of methane by biological reactions.

Methanogenic Bacterium : Bacteria that produce methane as a by: product of their chemical reactions.

Methanotroph : An organism capable of oxidising methane.

Micro : Prefix meaning small.

Microaerophile : Microorganisms that grow well in relatively low oxygen concentration environments.

Microaggregate : Clusters of clay stabilised by organic matter and precipitated inorganic matter.

Microarray : Sets of miniaturised chemical reaction areas that may also be used to test DNA fragments, antibodies, or proteins.

Microbial Biomass : Total mass of microorganisms living in a given mass or volume of soil.

Microbial Genetics : The study of genes and gene function in bacteria, archaea, and other microorganisms. Often used in research in the fields of bioremediation, alternative energy, and disease prevention.

Microbial Population : Total number of microorganisms living in a given mass or volume of soil.

Microbiology : The study of microorganisms, often with the aid of a microscope.

Microclimate : The climate of a very small area. For example, a single backyard can be considered a microclimate.

Microcosm : A community or any other unit that is representative of a larger community.

Micro–environment : The immediate physical and chemical surroundings of a microorganism.

Microfauna : Protozoa, nematodes and anthropods that are smaller than 200 microns.

Microflora : This includes bacteria, virus, fungi and algae.

Microgametophyte : The gametophyte developing vegetatively from the microspore of a heterosporous plant, male.

Microinjection : A technique for introducing a solution of DNA into a cell using a fine microcapillary pipet.

Micrometer : One millionth of a meter (10-6 meters).

Micronuclei : Chromosome fragments that are not incorporated into the nucleus at cell division.

Micronutrient : Elements that are required for growth in trace amounts. These include copper, iron, zinc etc.

Micro-organism : An organism that is too small to be seen by the naked eye. Also called microbes, these include bacteria, fungi, protozoans, algae and viruses.

Microphyll : A kind of leaf, specifically one which has a single, unbranched vein in it. Microphylls are only found in the lycophytes.

Microphyllous : Having small leaves that are usually hard and narrow.

Micropore : A small sized soil pore (approximately less than 30 microns in diameter) which is normally found within structural aggregates.

Micropropagation : Plant propagation from single cells under artificial conditions as created in the laboratory.

Micropyle : The small canal through the integuments (outer layers of tissue) of an ovule, usually at the point furthest away from the funicle (ovule stalk), persisting as a pore in the seed coat.

Microsite : A small part of the soil where the biological or chemical processes are different from the rest of the soil.

Microspecies : Segregate species of a larger species or species-aggregate.

Microsporangium : The sporangium in heterosporous plants where the cicrospores develop (pl. microsporangia).

Microspore : In plants which are heterosporous, the smaller kind of spore is called a microspore; it usually germinates into a male (sperm-producing) gametophyte.

Microsporocarp : A body containing the microsporangium, e.g. as in some ferns.

Microsporocyte : Diploid cells which on completion of meiosis produce microspores.

Microsporophyll : Specialised leaf bearing or subtending one or more microsporangia.

Microthallus : The tiny, inconspicuous stage of an organism's life history, alternating with the macrothallus stage.

Microtubule : Single proteinaceous tube like structure situated mostly in the plasma membrane. It regulates cellulose addition to the wall of the plant cells.

Middle Lamella : Layer of adhesive substance rich in pectin, which cements the cell walls of adjacent cells of multicellular plants together.

Midrib : Term applied to the mid vein or primary vein, especially when it is prominently raised or depressed.

Midvein : The primary vein which runs from the base to the apex of the lamina, usually the most prominent vein, from which arise the secondary or lateral veins.

Minor Veins : Small veins present on the leaf that branch off the lateral veins.

Minute : Very small, usually less than 1 mm long.

Mitochondrial DNA : The genetic material found in mitochondria, the organelles that generate energy for the cell. Not inherited in the same fashion as nucleic DNA.

Mitochondrion : Rod shaped organelles present in several eukaryotic cells, that work as powerhouse of the cell, by breaking down oxygen and nutrients and releasing energy in the form of ATP.

Mitosis : Nuclear division in which nuclear chromosomal material is initially duplicated and then split into two equal portions. Each daughter nuclei receives one portion of the nuclear chromosomes, thereby producing two genetically identical daughter nuclei.

Mixed Sporangia : Sporangia of all ages borne at all levels in sorus.

Mixotroph : Organisms that are capable of assimilating organic compounds as carbon sources, while using inorganic compounds as electron donors.

Model Organisms : A laboratory plant or other organism useful for research.

Modelling : The use of statistical analysis, computer analysis, or model organisms to predict outcomes of research.

Mold : A group of saprobic or parasitic fungi causing a cottony growth on organic substances.

Mole : The number of grams of a substance that equals its molecular weight in Daltons. For example, carbon has a molecular weight of about 12, thus 1 mole of carbon equals approximately 12 grams; water has a molecular weight of about 18, so 1 mole of water equals approximately 18 grams.

Molecular Biology : The study of the structure, function, and make-up of biologically important molecules.

Molecular Farming : The development of transgenic organism to produce human proteins for medical use.

Molecular Genetics : The study of macromolecules important in biological inheritance.

Molecular Medicine : The treatment of injury or disease at the molecular level. Examples include the use of DNA-based diagnostic tests or medicine derived from DNA sequence information.

Molecular Pump : Protein embedded in the membrane that forces molecules to pass from one side to another with the help of energy.

Monadelphous : Of stamens, with their filaments fused into one group, as in many Malvaceae.

Moneocious : (one house) producing male and female gametes on the same individual.

Moniliform : Shaped like a cylindrical body which is constricted at regular intervals; resembling a string of pearls.

Mono : Prefix meaning 'one' or 'solitary' generally bean-shaped; also referred as 'bilateral'.

Monoclonal Antibody : Antibody produced from a single clone of cells, which has a uniform structure and specificality.

Monocot : One of two main divisions of flowering plants, characterised by having a single cotyledon (seed leaf); examples include grasses, orchids, bamboos, and palms.

Monocotyledons : A major group of angiosperms, characterised by the embryo usually having one cotyledon.

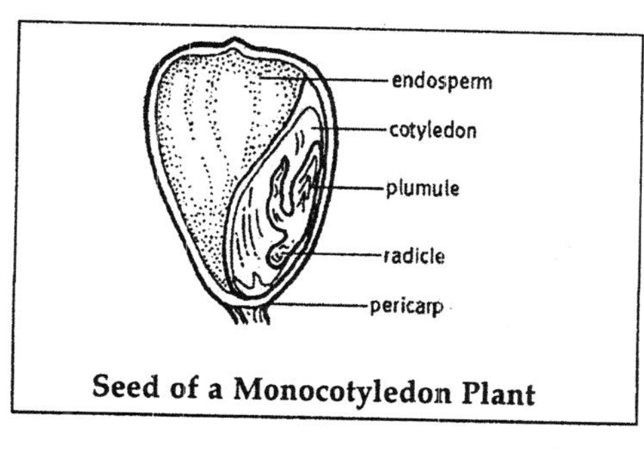

Seed of a Monocotyledon Plant

Monoecious : Plants which possess both unisexual male and female flowers or cones on the same plant.

Monogenic Disorder : A disorder caused by mutation of a single gene.

Monokaryon : Fungal hyphae where the compartments contain only nucleus.

Monophyletic : Derived from a single ancestral line.

Monopodial : Mode of development where primary axis maintains main line of growth and secondary laterals are produced off that main axis.

Monopodium : An axis in which the growth is continued from year to year by the same growing point; hence monopodial.

Monosomy : Possessing only one copy of a particular chromosome instead of the normal two copies.

Monostromatic : Blade composed of a single layer of cells.

Monotypic : Having only one representative, eg a genus or family with a single species.

Montane : Of or pertaining to, or growing in, the mountains.

Morbid Map : A diagram showing the chromosomal location of genes associated with disease.

Morphic : Suffix pertaining to 'form', eg. dimorphic, polymorphic etc.

Morphology : The study of form and structure of organisms.

Morphometric Characters : These are characteristics regarding the depth, dimension, sediment distribution, water currents etc.

Motile : Actively moving, self-propelled.

Motility : The ability of a cell to move from one place to another.

Mucigel : Gelatinous material found on the surface of roots growing in normal soil.

Mucilage : Gelatinous secretions and exudates produced by plant roots and most microorganisms.

Mucilaginous : Slimy or pertaining to mucilage.

Mucro : A short, sharp tip, such as a prolong midrib; hence mucronate.

Mucronate : Having a short projection at the tip, as of a leaf.

Mulch : Materials which are laid down on soil to protect it from rain, crusting, freezing etc. these materials could be sawdust, plastic, leaves etc.

Multi : Prefix meaning 'many'.

Multialxial : Axis composed of multiple, longitudinal filaments, each derived from one apical cell.

Multifarious : Many-ranked, in many rows.

Multifid : Cleft into very many narrow lobes or segments.

Multifoliate : Bearing many leaves.

Multiple Fruit : A cluster of fruits produced from more than one flower and appearing as a single fruit, often on a swollen axis, as in Moraceae.

Multiple Fruit

Multiplexing : A laboratory approach that performs multiple sets of reactions in parallel (simultaneously); greatly increasing speed and throughput.

Municipal Solid Waste : The total consumer and commercial waste generated in a certain confined and restricted geographic area.

Muricate : Of a surface, rough with pointed protuberances or short hard tubercles.

Murine : Organism in the genus Mus. A rat or mouse.

Mutagen : An agent that causes a permanent genetic change in a cell. Does not include changes occurring during normal genetic recombination.

Mutagenicity : The capacity of a chemical or physical agent to cause permanent genetic alterations.

Mutation : Any heritable change in DNA sequence.

Muticous : Blunt; lacking a distinct process.

Mycophagous : Organisms that eat fungi.

Mycorrhiza : A fungus associated with the root of a plant, the association of mutual benefit; hence mycorrhizal.

Mycorrhizal : Having a symbiotic relationship between a fungus and the root of a plant.

Mycovirus : Viruses that infect fungi.

Myrmecophilous : Ant-loving, of plants inhabited by ants and offering specialised shelters and food for them.

Mysticetes : Whales of the suborder Mysticeti, like Right whales, finback, gray whale, humpback whales, rorquals, etc.

N

Nacreous : Having a pearly luster.

Naked : Of flowers, without a perianth; of sporangia, not covered with an indusium; of seeds, not enclosed in an ovary, exposed on the surface of a sporophyll.

Naked DNA : Gene transfer processes such as transformation and transfection involves the passage of nucleosome and histone free DNA. This histone and nucleosome free DNA is called naked DNA.

Nanoplankton : Plankton from 2 to 20 microns in diameter that would pass through the mesh of a normal plankton net.

Nanopore : Soil pore having dimensions in nanometers.

Nascent : In the process of being formed.

Nastic Movement : Non-directional movement of flat plant organs such as leaf, petal, etc. irrespective of the stimulus position.

Native : Naturally occurring in the area, but not necessarily confined to it.

Natural Selection : The process of evolution involving the population rise of organisms which have inherited the traits that enable them to successfully survive in natural conditions and reproduce successfully in comparison to others.

Naturalised : Originating elsewhere but established and reproducing itself as though native to the area.

Necrosis : Death of plant cells or tissues, leading to discoloration of leaves and stems. It can even conduce death of the plant.

Necrotroph : Fungus which attacks the host in a virulent manner, and then kills it. They then absorb all the nutrients from the dead organism.

Necrotrophic : A mechanism by which an organism produces lytic enzymes that kill and then breakdown host cells for its nutrition.

Nectar : A more or less sweet fluid secreted from a specialised gland or nectary.

Nectary : Gland secreting a sweet fluid (nectar) commonly in insect-pollinated flowers but not restricted there.

Nemathecium : Wartlike structure bearing reproductive parts.

Nematocyst : This refers to tiny hairlike structures in coelenterates which is used by them to eject stingers.

Neonate : The phenomenon of producing live young ones instead of laying eggs.

Neotrophics : The tropical areas of the 'New World', central America and the northern part of south America.

Neotype : A specimen selected to serve as the nomenclatural type as long as the original material is missing.

Nephridium : A tube like excretory organ of many invertebrates such as mollusks and earthworms.

Neritic : Living in coastal ocean waters.

Net Veins (Reticulate) : Forming a network or reticulum; e.g. of veins.

Netted : Same as reticulated, in the form or pattern of a network.

Neuston : Organisms that live at the ocean-atmosphere interface.

Neuter : Lacking a pistil or stamens.

Neutralism : Lack of interaction between two organisms in the same habitat.

Niche : Functional role of an organism in a certain habitat.

Nicotinamide Adenine Dinucleotide Phosphate (NADP+) : An important oxidised co-enzyme that acts as a hydrogen and electron carrier in various redox reactions.

Nictotinamide Adenine Dinucleotide (NAD+) : An important oxidised co-enzyme that is a hydrogen and electron carrier in redox reactions.

Nidicolous : The time spent in the nest after its hatched.

Nidifugous : The phenomenon of leaving the nest within a few days of hatching.

Nidulent : Lying within a cavity, embedded within a pulp.

Nitrate Reduction (Biological) : The process of reduction of nitrate to simpler forms like ammonium by plant and microorganisms.

Nitrification : Biological oxidation of ammonium to nitrite and nitrate.

Nitrifying Bacteria : Chemolithotrophs that can carry out the transformation from ammonia to nitrite or nitrate.

Nitrogen (N) : Chemical used by plants to synthesise proteins, enzymes, and chlorophylls. When a plant is not able to get enough nitrogen, leaves tend to yellow from the bottom of the plant upwards, and from the leaf tip to the stem.

Nitrogen Assimilation : Process of ammonium incorporation into organic compounds present within an organism.

Nitrogen Cycle : The cycle where nitrogen is used by a living organism, then after the organism dies is restored to soil, followed by its final conversion to its original state of oxidation.

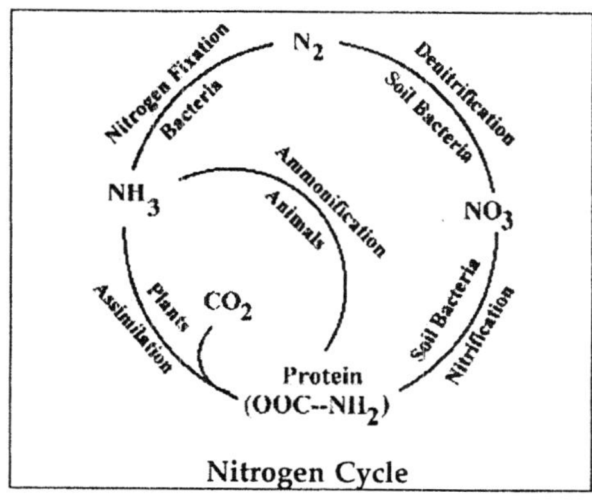

Nitrogen Cycle

Nitrogen Fixation : Process by which plants convert atmospheric nitrogen into compounds such as nitrate or ammonium, which they can readily use.

Nitrogenase : The enzyme required for biological nitrogen fixation.

Node : The region of a stem between two inter-nodes, where there is branching of the vascular tissue into leaves or other appendages.

Nomen Conservandum : The name of a taxon internationally agreed upon to be exempt from the rules of priority of publication.

Nomenclatural Synonyms : Synonyms (viz.) with the same basionym (viz.) or based on the same type (viz.), the result of the transference of a species from one genus to another.

Nomenclature : The study of the application of the names of taxa.

Nomophilous : Growing in or loving pastures.

Nonpolar : A substance that is hydrophobic and does not easily dissolve in water.

Northern Blot : A gel-based laboratory procedure that locates mRNA sequences on a gel that are complementary to a piece of DNA used as a probe.

Notate : Marked with lines or spots.

Nucellus : Central region of an ovule, wherein embryo sac development takes place.

Nuclear Envelope : The porous double lipid bilayer sheathing the nucleus.

Nuclear Transfer : A laboratory procedure in which a cell's nucleus is removed and placed into an oocyte with its own nucleus removed so the genetic information from the donor nucleus controls the resulting cell. Such cells can be induced to form embryos. This process was used to create the cloned sheep "Dolly".

Nucleic Acid : Macromolecule composed of chains of nucleotides, carrying genetic information.

Nucleoid : The nuclear region of certain organisms like bacteria, which contains chromosomes, but which is not limited by a nuclear membrane.

Nucleolar Organising Region : A part of the chromosome containing rRNA genes.

Nucleolus : Spherical structure which is non-membranous and comprises proteins and nucleic acids. It is present within the nucleus, and each nucleus may contain more than one nucleolus.

Nucleophilic Compound : An electron donor in chemical reactions involving covalent catalysis in which the donated electrons bond with other chemical groups.

Nucleotide : A subunit of DNA or RNA consisting of a nitrogenous base (adenine, guanine, thymine, or cytosine in DNA; adenine, guanine, uracil, or cytosine in RNA), a phosphate molecule, and a sugar molecule (deoxyribose in DNA and ribose in RNA). Thousands of nucleotides are linked to form a DNA or RNA molecule.

Nucleus : Largest cell organelle found in most eukaryotic cells. It contains most of the cell's genetic material, thus is involved with inheritance, ribosome synthesis and metabolism control.

Numerous : Eleven or more, same as 'many'

Nut : A dry fruit consisting of only one seed and a thick pericarp. Nuts feature a cluster of bracts at their base, e.g., hazelnut, acorn.

Nutlet : A small nut or one of the sections of the mature ovary of some members of the *Boraginaceae, Verbenaceae* or *Lamiaceae.*

Nutritive Cell : Specific cell of carpogonial branch with which carpogonium fuses after fertilisation in some red algae of the order Cryptonemiales.

O

Ob : Prefix meaning 'inverted', eg. obconic, oblanceolate, obcordate.

Obconic (-al) : Inversely conic (-al), a cone attached at the narrow end.

Obconic : Inversely cone-shaped and attached at the pointed end.

Obcordate : Inversely heart-shaped, attached at the point.

Oblanceolate : Of a plane several times longer than wide, widest in the apical third, tapering gradually towards the base, more rapidly towards the apex.

Oblate : Almost circular, but with breadth slightly greater than the length.

Obligate : Is an adjective that means "necessary" when used in biology. Also, exhibited by all members of a species without exception.

Obligate Aerobe : An organism that requires air for aerobic cellular respiration. They use oxygen in order to oxidise substrates and obtain energy.

Oblique : Of a leaf, leaf base or other organ, having the sides unequal.

Oblong : Two to four times longer than broad with nearly parallel sides, but broader than 'linear'

Obovate : Inversely ovate, of a plane with the outline of an egg, attached at the narrower end.

Obovoid : Inversely ovoid, of a solid with the form of an egg, attached at the narrower end.

Obpyriform : Pear-shaped but broadest above the middle.

Obsolescent : Not functional, but not reduced to a rudiment.

Obsolete : Reduced to a rudiment, or completely lacking.

Obtuse : Blunt or broadly rounded, the converging edges separated by an angle greater than 900, e.g. of an lamina apex.

Obverse : Describing a leaf that is narrower at the base than at the apex.

Obvolute : A vernation in which two leaves are overlapping in the bud in such a manner that one-half of each is external and the other half is internal, i.e. each leaf both overlaps the next and is in turn overlapped by the one before.

Ochraceus : Ochre-coloured; dull yellow with a tinge of red.

Ochreoleucous : Yellowish-white, cream-coloured.

Ocrea : A sheath around the stem derived from the leaf stipules, primarily used in the *Polygonaceae.*

Octo : Prefix meaning eight.

Odd-pinnate : Describing a pinnately-compound leaf with a single terminal leaflet .

Odontocetes : Used while referring to any whale of the suborder Odontoceti, such as killer whales, dolphins and sperm whales. They are characterised by a single blowhole, an asymmetrical skull and rows of teeth. They feed primarily on squid, fish and crustaceans.

Offset : A young plant, developing as a vegetative offshoot from the parent plant.

Oil Glands : Small structures embedded in a leaf or other organ, secreting a volatile oil, mostly visible as small translucent dots (hand lens needed) against a strong light; usually making the organ aromatic when crushed.

Oligomeris : With less than the typical number of parts.

Oligonucleotide : A short nucleic acid chain, which is obtained from an organism or is synthesised chemically.

Oligotroph : A microorganism that has adapted itself to grow in environments that are low in nutrients.

Olivaceous : Olive-coloured, a yellowish green darkened with black.

Oncogene : A gene, one or more forms of which is associated with cancer. Many oncogenes are involved, directly or indirectly, in controlling the rate of cell growth.

One-foliolate (1-Foliolate, Unifoliolate) : A compound leaf reduced to a single leaflet, usually recognised by the articulated or jointed 'petiole', which is in fact a petiole plus a petiolule.

Ontogeny : The development of a single organism, the stages through which it passes in its lifetime.

Oogamy : It is a kind of sexual reproduction where the female gamete is non motile and larger than the motile male gamete.

Oogonium : It is the term given to the female sex organ of various algae and certain fungi.

Open : Uncongested, usually describing the organisation of flowers in an inflorescence.

Open Forest : A forest dominated by trees with relatively narrow isobilateral leaves forming sparsely foliaged crowns (usually species of eucalypts); the forest canopy is sparse and often not continuous, allowing sunlight to reach the ground within the forest.

Open Reading Frame (ORF) : The sequence of DNA or RNA located between the start-code sequence (initiation codon) and the stop-code sequence (termination codon).

Open-pollinated Plant : A horticultural term used to describe cultivars of a plant that are bred through natural, random pollination.

Operculum (calyptra) : A cap-like covering or lid of some flowers or fruits that becomes detached at maturity by abscission; e.g. (1) the cap on the buds of eucalypts, (2) the lid of circumsciss capsules.

Operon : A set of genes transcribed under the control of an operator gene.

Opposite : Said of leaves, etc. when they arise on opposite sides of a stem at the same node; of flower parts when they are situated directly above other parts.

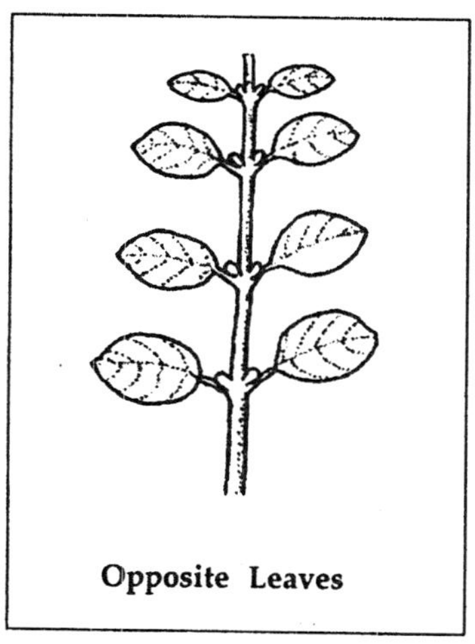

Opposite Leaves

Orbicular : Orb-shaped, strictly a 3-dimensional shape but often used for a 2.

Order : A taxonomic group consisting of one or more closely related families.

Organelle : The membrane bound cell bodies found in the cytoplasm. Eg : Mitochondria.

Organic Matter : Any material that originates from a living organism.

Organism : A complete, living individual of a species.

Orifice : A small opening or aperture.

Orophilous : Growing in or preferring mountain areas.

Ortho : Prefix meaning 'straight'.

Osmosis : The differential behaviour of membrane for the purpose of diffusion of water and other solvents. Osmosis always happens from the region of higher concentration to lower concentration.

Osmotic Potential : This is the minimum pressure required to prevent osmosis from taking place. It is the pressure developed by a solution that is separated from water by a selectively permeable membrane.

Ospore : Thick: walled spore formed in an oogonium by fungus like organisms like the phylum Oomycota.

Osteichthyes : Are a taxonomic group of fish that includes the lobe-finned fish (Sarcopterygii) and ray-finned fish (Actinopterygii). They are also referred to as bony fish.

Ostiole : Opening in conceptacle or cystocarp through which spores or gametes can escape.

Ostracum : The calcified portion of an invertebrate's shell. While the organism is living, the ostracum is covered by layers of protein forming a periostracum.

Outcrossing : The pollination which takes place between two different flowers which may or may not belong to the same genetic line.

Oval : Broadly elliptic, narrowing somewhat from the middle to rounded ends.

Ovary : The basal portion of a carpel or a group of fused carpels in which one or more ovules are enclosed, and which after fertilisation develops into the fruit.

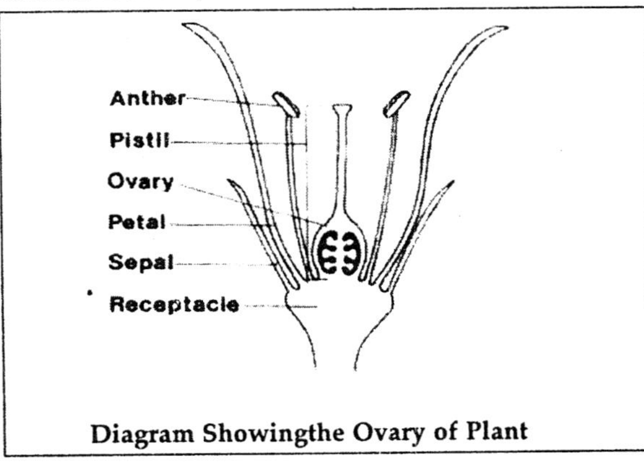

Diagram Showingthe Ovary of Plant

Ovate : A 2-dimensional shape, with the length 13 times the breadth, and broadest below the middle.

Overtopping : The ability of a shoot to grow for longer period of time than the other shoot in the same plant, which was a result of branching.

Ovoid : Of a solid with the form of an egg, attached at the broad end.

Ovule : A structure in seed plants within which one or more megaspores are formed, and which after fertilisation develops into the seed.

Oxidation Phosphorylation : The energy released during metabolic pathways, which is responsible for the formation of ATP and ADP.

P

Palea **:** (1) the upper of 2 bracts enclosing the flower of a grass; (2) one of the chaffy scales on the petiole and rachis of many ferns.

Paleo Herb **:** Any member of a group of basal flowering herbs which may be the closest relatives of the monocots. They include the water lilies, Piperales, and Aristolochiales.

Paleotropics **:** The tropical areas of the 'Old World', Africa, SE Asia and the western Pacific.

Palindrome **:** This term refers to a DNA sequence which can be read forward or backward.

Palisade Mesophyll **:** It is also known as palisade or palisade parenchyma, and is the upper layer of ground tissue in a plant leaf. It comprises of prolonged cells underneath and vertical to the upper cuticle, and constituting the principal area of the photosynthesis process.

Palmate **:** (1) of a compound leaf with 3 or more leaflets arising from the one point at the top of the petiole, (2) of veins in a lamina, radiating from the one point, adv. palmately.

Palmately Compound **:** Used to describe a leaf whose leaflets arise from a common point.

Palmately Trifoliolate **:** Of a leaf, with 3 leaflets arranged palmately, i.e. all the petiolules of about the same length.

Palmatifid **:** Of a leaf cut into lobes to less than halfway in a palmate form.

Palmatisect : Of a leaf cut into lobes to more than halfway in a palmate form.

Paludose : Growing in wet meadows or marshes.

Pandurate : Fiddle-shaped, +/- obovate but with a waist.

Panicle : A compound inflorescence with a main axis and lateral branches which are further branched, and in which each axis ends in a flower or flower bud. adj. paniculate.

Pannose : With a covering of short, dense, felty or woolly tomentum.

Pantropic(-al) : Found throughout the tropics.

Papilionaceous : Describing the structure of a corolla typical of the *Fabaceae* with banner, wings and keel.

Pappus : The group of appendages, usually hairs or scales, above the ovary and outside the corolla in Asteraceae (possibly a modified calyx); often persisting on the fruit and aiding in its dispersal.

Parallel : Of veins in a lamina, all running in the same direction and equally distant from one another, as in grass leaves.

Paramylon : A glucose polymer which serves a food reserve in Euglenophyta.

Paraphysis : A sterile hairlike structure among sporangia or gametangia, as those in the brown algal order Fucales.

Parasexual Cycle : A nuclear cycle wherein genes of haploid nuclei recombine without meiosis.

Parasite : An organism growing and feeding upon another organism (the host). A hemiparasite is partly parasitic, partly autotrophic, and has some chlorophyll.

Parasitism : Feeding by one organism on the cells of a second, normally larger organism, thus, harming the host.

Paratype : Specimens cited at the same time as the original description, other the holotypes and isotypes.

Parenchyma : A generalised cell or tissue in a plant. These cells may manufacture or store food, and can often divide or differentiate into other kinds of cells.

Parietal : (1) attached to the wall; (2) of placentation, with placentas on the wall or intruding partitions of a unilocular compound ovary.

Paripinnate : Term describing a pinnately compound leaf without a single terminal leaflet, and therefore usually with an even number of leaflets.

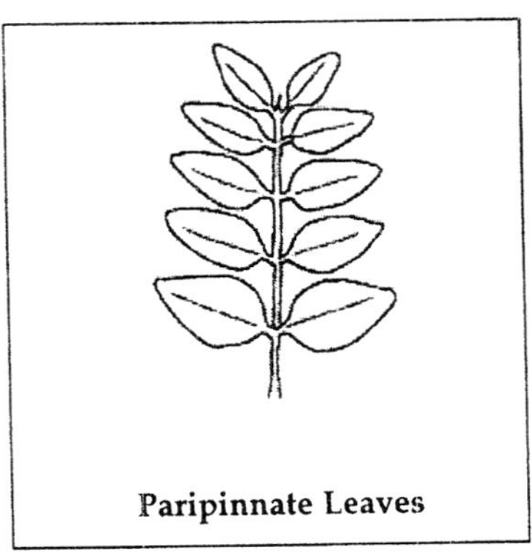

Paripinnate Leaves

Parted : Lobed or cut in over halfway and often very close to the base or midrib.

Parthenocarpy : It is the activity wherein a fruit is produced without egg fertilisation in the ovary.

Parthenogenetic : Production of new individual from single, unfertilised gamete, usually the egg.

Partial Inflorescence : A portion of an inflorescence, particularly a part sufficiently branched to show the same structural plan as the inflorescence as a whole.

Particle Density : Density of particles present in soil.

Particle Size : Effective diameter of a particle measured by sedimentation or micrometric methods.

Partite : Suffix meaning 'deeply divided to the base or almost so, into equal parts', eg bipartite, tripartite, 4-partite etc.

Pasteurisation : Process of using heat to kill or reduce the activity of microorganisms in heat: sensitive materials.

Patelliform : Knee-shaped; shaped like a small dish, circular and rimmed.

Pathogen : An organism that is capable of causing an infection, or harming a host cell.

Pathogen Suppressive Soil : Soil where a pathogen does not persist, either in its own survival or in its pathogenicity.

Pathogenicity : The ability of a parasite to infect or inflict damage on a host.

Peat : Unconsolidated soil material consisting mostly of undecomposed organic matter with excessive moisture content.

Pectinate : Describing a pinnatifid leaf whose segments are narrow and arranged like the teeth of a comb.

Pedate : Term describing a palmately compound leaf with the lateral leaflets divided again.

Pedicel : The stalk of an individual flower in an inflorescence; the stalk of a sporangium or of a conceptacle; hence pedicellate.

Pedigree : A family tree diagram that shows how a particular genetic trait or disease has been inherited.

Peduncle : The stalk of a flower cluster, or of a solitary flower not associated with others in an inflorescence.

Peep : A generic name for several sandpiper species.

Pelagic : Organism that live and thrive in open oceans or seas rather than waters adjacent to land.

Pellicle : A rigid protein layer just below the cell membrane.

Peltate : Term describing an organ with a stalk or point of attachment on its lower surface away from the margin, often umbrellalike; e.g. leaves, scales.

Pendent : Hanging downward. Also spelled pendant.

Penetrance : The probability of a gene or genetic trait being expressed. "Complete" penetrance means the gene or genes for a trait are expressed in all the population who have the genes. "Incomplete" penetrance means the genetic trait is expressed in only part of the population. The percent penetrance also may change with the age range of the population.

Penniveined, -veined : Pinnately arranged veins in which the secondary veins are conspicuous and numerous and are more or less parallel to each other, as in a feather.

Penta : Prefix meaning 'five'.

Pentaradial Symmetry : The balanced distribution of duplicate body parts or shapes (sensory and feeding structures) in a five fold circular pattern, i.e body parts arranged in fives or multiples of 5, in a symmetry in organisms.

Pepo : A fleshy, indehiscent fruit with a hard, more or less thickened rind and a single many-seeded locule, characteristic of the Cucurbitaceae.

Peptide : Two or more amino acids joined by a bond called a "peptide bond.".

Peptidoglycan : Rigid cell wall layer seen in bacteria. It's also called murein.

Perenate : Maintain a dormant state through the non-growing season; hence perennating buds etc.

Perennial : A plant which continues to grow after it has reproduced, usually meaning that it lives for several years.

Perfect : Containing both stamens and pistils.

Perfoliate : Of a leaf or bract completely encircling the axis and so the stem apparently passing through it.

Perianth : The calyx and corolla collectively; especially when they are similar, individual segments then being called tepals.

Peribacteroid Membrane : A plant derived membrane which surrounds rhizobia in host cells of legume nodules.

Pericarp : It relates to the matured and diversely altered walls of a plant ovary.

Pericycle : It is the outermost cell layer of the stele in a plant, which often turns into a zone that is multilayered.

Periderm : This term pertains to the bark, and comprises of cork, cork cambium, and any enclosed tissues like secondary phloem.

Perigynous : Term describing sepals, petals and stamens that are attached to the rim of a lateral or upward expansion of the receptacle, or attached to the rim of a hypanthium which is not fused to the ovary.

Periostracum : The external, outermost covering of the shell of some mollusks. It helps to protect the slender slimy inner portions as well as provide the shell with colour.

Periplasmic Space : The area between the cell membrane and cell wall in Gram negative bacteria.

Perisperm : Nutritive tissue within the seed which is formed outside the embryo sac, lies between the embryo sac and the testa.

Perispore : A wrinkled or folded outer covering to some spores (also epispore).

Peristome : A set of cells or cell parts which surround the opening of a moss sporangium. In many mosses, they are sensitive to humidity, and will alter their shape to aid in spore dispersal.

Perithecium : Flask shaped ascocarp open at the tip.

Peritrichous Flagellation : Multiple flagella present all over the cell surface.

Permanent Wilting Point : The highest concentration of soil at which plants present in it, will irreversibly wilt when placed in a humid chamber.

Persistent : Remaining until the part that bears it is fully matured, e.g. of floral parts remaining until fruit is mature; of a leaf base, remaining attached to the plant after the leaf or frond has been shed.

Petal : One of the outer appendages of a flower, located between the outer sepals and the stamens. Petals often display bright colours that serve to attract pollinators.

Petaloid : Resembling a petal, especially in colour and texture.

Petiole : Portion of the leaf which joins the leaf to the node of a stem.

PH **:** A measure of acidity and basicity. A pH of 7 is defined as neutral. Lower pH values (e.g. 3-4) correspond with a greater acidity. Higher pH values (e.g. 9-10) correspond with a higher alkalinity (or basicity).

Phage **:** A virus for which the natural host is a bacterial cell.

Phagotrophy **:** Endocytosis or engulfing particles of food as a mode of nutrition.

Phanerogam **:** Seed-plant or spermatophyte, eg. the flowering plants and gymnosperms.

Pharmacogenomics **:** The study of the interaction of an individual's genetic make-up and response to a drug.

Phenocopy **:** A trait not caused by inheritance of a gene but appears to be identical to a genetic trait.

Phenology **:** The study of flowering or fruiting periodicity of plants.

Phenotype **:** The physical characteristics of an organism as opposed to its genetic composition or genotype; hence phonetic.

Phil, Philous **:** Suffix meaning 'liking' or 'preferring'.

Phloem **:** One of two types of vascular tissue (the other being xylem), which transports nutrients produced in the leaves downwards to the root system, as well as to nonphotosynthetic portions of the shoot system.

Phosphobacterium **:** Bacteria that are good at dissolving insoluble inorganic phosphate that is present in soil.

Phosphorous (P) **:** Chemical involved in flowering, fruiting, root growth, and energy transfer throughout the plant. Plants short in phosphorous will show stunted growth.

Photoautotroph **:** An organism which uses light energy to synthesise organic compounds from CO_2.

Photoheterotroph **:** Organisms able to use light as source of energy and organic materials as carbon source.

Photoperiodism **:** Physiological response due to length of light exposure.

Photophosphorylation : Synthesis of high energy phosphate bonds by the use of light as source of energy.

Photorespiration : Metabolic pathway that uses oxygen and produces CO_2.

Photosynthesis : Photosynthesis is a plant activity which includes the synthesis of complex organic substances, peculiarly saccharides, from carbon dioxide, water, and inorganic salts, utilising sunlight as a source of energy and with the help of chlorophyll and associated pigments.

Photosystem I and II : Photosystem I absorbs light for the transfer of negatron from plastocyanin to ferredoxin. Its reaction centre is P700. Photosystem II absorbs light for oxidation of water and reduction of plastoquinone. Its reaction centre is P680.

Phototaxis : Movement of an organism, or a part of it, towards light.

Phragmoplast : Microfibril parallel to the spindle axis at telophase across which a cell plate is deposited in cell division.

Phreatophyte : A perennial plant that has deep and extensive root systems that enable it to tap underground sources of water.

Phycobilin : Water soluble pigment that is seen in cyanobacteria and is the light harvesting pigment for Photosystem II.

Phycobilisome : Organelle on surface of thylakoids in which biliprotein pigments are present in Cyanobacteria and Rhodophyta.

Phycocyanin : Blue water-soluble biliprotein pigment of Cyanobacteria and Rhodophyta.

Phycoerythrin : Red water-soluble biliprotein pigment of cyanobacteria and Rhodophyta.

Phycoplast : Assemblage of microtubules perpendicular to the spindle and at the equator of the cell at telophase in cell division.

Phyll, Phyllo : Suffix or prefix meaning 'leaf'.

Phyllary : An involucral bract of the Asteraceae, collectively the phyllaries surrounding a head form the calyculus.

Phyllode : A flattened petiole, leaflike in appearance and function, replacing the lamina, as in many wattles.

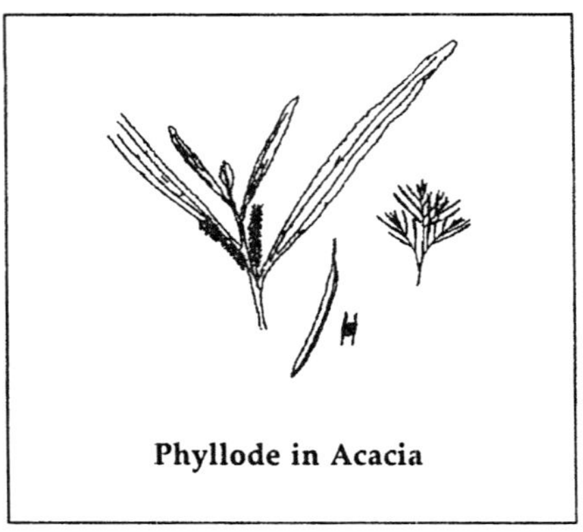

Phyllode in Acacia

Phyllopodium : An outgrowth of the rhizome in ferns to which the frond or stipe is joined.

Phyllotaxis, Phylotaxy : The arrangement of leaves etc. around an axis.

Phylogeny : The evolutionary development of a group and its derivation from ancestors and the relationship between its members; hence phylogenetic.

Physical Map : A map of the locations of identifiable landmarks on DNA (e.g., restriction-enzyme cutting sites, genes), regardless of inheritance. Distance is measured in base pairs. For the human genome, the lowest-resolution physical map is the banding patterns on the 24 different chromosomes; the highest-resolution map is the complete nucleotide sequence of the chromosomes.

Phytochrome : This term pertains to a plant pigment which is involved in the soaking up of light in the photoperiodic response which modulates various types of growth and development.

Phytoextraction : The use of plants or algae for removing contaminants from soil, sediments or water, and turning them into harvestable plant biomass.

Phytomelanin : A papery "sooty" black layer over the seed of plants in the Asparagales, which includes agaves, aloes, onions and hyacinths. It is an important character for defining the group.

Pilose : Hairy with long soft weak hairs which are clearly separated but not sparse.

Pilus : Fimbria like substance present on fertile cells that deals with transfer of DNA during the process of conjugation.

Pinna : The leaflet of a pinnate leaf, or the primary division of a bi- or tripinnate leaf (dimin. pinnule).

Pinnate : (1) (1-pinnate) of a leaf, with the lamina divided into pinnae in 2 rows along a rachis, once compound, (2) of veins, with the secondary veins arranged regularly.

Pinnate Compound Leaf : Compound with the leaflets arranged on each side of the rachis.

Pinnately Trifoliolate : Of a leaf, with three leaflets arranged pinnately, i.e. the terminal petiolule usually jointed and longer than the lateral ones.

Pinnatifid : Of simple leaves or leaflets, of the lamina cut into lobes on both sides of the midrib.

Pinnatipartite : Pinnately lobed half to two thirds the depth of the lamina.

Pinnatisect : Of simple leaves or leaflets, of the lamina cut down almost to the midrib but having the segments confluent with it. e.g. as in the ultimate segments of some fern fronds.

Pinnule : The second or third order divisions of a bi- or tripinnate leaf; the ultimate free divisions of such a leaf.

Pinnate : Having branches, lobes, leaflets or veins arranged on opposite sides of a common axis.

Pistil : The female organ of a flower which bears ovules or seeds, consisting of a complete ovary, style, and stigma.

Pistillate : A female flower that has two or more pistils but no functional stamens.

Pistillate Flower : A flower with pistils; without stamens.

Pit : It is concerned with the portion of a sclerenchyma cell, where there is no secondary wall over the primary one, and substances are able to pass into or out of the cell.

Pit Connection : Discrete len-shaped plug of cytoplasm held between two adjacent cell walls in the Rhodophyta.

Pith : This is a soft and squishy central cylinder of parenchymatous tissues in the stalks of plants having two cotyledons in the seed.

Pitted : Having numerous small depressions in the surface.

Placenta : The part of the ovary to which the ovules are attached.

Placentation : The arrangement of the placentas and the attached ovules.

Plane : With a flat, smooth surface.

Plankton : Microscopic organisms like algae and protozoa that drift on the oceans' currents.

Planoconvex : With upper surface convex, lower surface more or less flat.

Plant Anatomy : Study of the internal structure of the plant.

Plant Geography : It is also known as phytogeography, phytochorology, geobotany, geographical *botany*, or vegetation science; and refers to spacial distribution of plants and vegetation in different environments and regions.

Plant Physiology : The study of plants, which involves processes such as nutrition, reproduction, and other functions.

Plant Taxonomy : The science that refers to the identifications, description, naming, and classification of plants according to their unique characteristics.

Plaque : A localised area of lysis or cell inhibition which is caused due to virus infection.

Plasmid : Autonomously replicating extrachromosomal circular DNA molecules, distinct from the normal bacterial genome and nonessential for cell survival under nonselective conditions. Some plasmids are capable of integrating into the host genome. A number of artificially constructed plasmids are used as cloning vectors.

Plasmodesmata : It pertains to a narrow hole in the elementary wall, that comprises of some cytol, cell membrane, and a desmotubule. It is a means of communication between cells.

Plasmodium : Body of slime mold, which is a large mass of living substance with hundreds or thousands of karyons. Plasmodium ingest fungal spores, bacteria and other tiny protozoans.

Plasmogamy : Fusion of two cell contents, inclusive of the cytoplasm and nuclei.

Plasmolysis : It is an activity which relates to the shrinking of the living substances when water is removed by exosmosis.

Plastic : Influenced in form by the environment.

Plastid : These are major organelles found in plant cells, as well as algal cells. These organelles are sites of manufacture of various essential chemical compounds used by the cell. They often contain chlorophyll, which is used for photosynthesis.

Plastron : The ventral surface of the shell of a tortoise or a turtle.

Plate Count : Number of colonies formed on a solid culture medium, when uniformly inoculated with a known amount of soil.

Platyspermic : Having seeds which are flattened and disc-like.

Plectostele : A complex stele with the vascular bundles arranged in several parallel ribbons.

Pleiotropy : One gene that causes many different physical traits such as multiple disease symptoms.

Plicate : Folded like a paper fan, as in the leaves of palms, cyclanthoids, and some orchids.

Plumose : Featherlike, with a central axis and fine hairs arising from it; e.g. (1) the styles in Clematis species, (2) the pappus of some Asteraceae.

Plumule : It is the bud of the plant axis which moves up while it is still in the embryo.

Pluri : Prefix meaning 'several'.

Pluricellular : Of a hair of several cells and opposed to unicellular.

Pluripotency : The potential of a cell to develop into more than one type of mature cell, depending on environment.

Pneumathodes : Bands or pores or aerating tissue, especially along the stipes of ferns.

Pneumatocyst : Gas bladder or float as in the brown alga *Macrocystis*.

Pneumatophore : It refers to a differentiated structure that originates from the root in particular plants which spring up in swamplands and fenlands, and act as a respiratory organ. Eg: Mangroove.

Pneumatophores : Specialised vertical roots produced by some vascular plants which grow in water, waterlogged mud or tidal swamps; the roots contain spongy tissue which enables them to exchange gases with the atmosphere through lenticels in their aerial portions, as in many mangroves.

Pod : A legume or superficially similar fruit.

Polar Flagellation : The presence of flagella at one or both ends.

Pollen : The microspores of seed plants, formed in and shed from the anthers, by which time some nuclear division has taken place to form a gametophyte enclosed in the spore wall.

Pollen Grains : They relate to microspores in seed plants, that comprise of a male gametophyte.

Pollen Tube : In seed plants, the extension of the male gametophyte as it emerges from the pollen grain in search of the female gametophyte.

Pollination : Process of transferring the pollen from its place of production to the place where the egg cell is produced. This may be accomplished by the use of wind, water, insects, birds, bats, or other means. Pollination is usually followed by fertilisation, in which sperm are released from the pollen grain to unite with the egg cell.

Pollinia : A mass of fused pollen produced by many orchids.

Pollinium : A cohered mass or body of pollen grains, characteristic of plants which belong to the orchid and milkweed families.

Poly : Prefix meaning many.

Polyandrous : With many stamens.

Polycephalous : With many flower heads.

Polygamous : Having both unisexual and bisexual flowers on the same plant.

Polygenic Disorder : Genetic disorder resulting from the combined action of alleles of more than one gene (e.g., heart disease, diabetes, and some cancers). Although such disorders are inherited, they depend on the simultaneous presence of several alleles; thus the hereditary patterns usually are more complex than those of single-gene disorders.

Polyhphyletic : Composed of members that descended independently from two or more ancestral lines.

Polymer : It is a big chemical compound that consist of several subunits called as monomers.

Polymerase Chain Reaction (PCR) : A method for amplifying a DNA base sequence using a heat-stable polymerase and two 20-base primers, one complementary to the (+) strand at one end of the sequence to be amplified and one complementary to the (-) strand at the other end. Because the newly synthesised DNA strands can subsequently serve as additional templates for the same primer sequences, successive rounds of primer annealing, strand elongation, and dissociation produce rapid and highly specific amplification of the desired sequence.

Polymerase, DNA or RNA : Enzyme that catalyses the synthesis of nucleic acids on preexisting nucleic acid templates, assembling RNA from ribonucleotides or DNA from deoxyribonucleotides.

Polymorphic (Polymorphous) : Existing in several or many forms.

Polymorphism : Difference in DNA sequence among individuals that may underlie differences in health. Genetic variations occurring in more than 1% of a population would be considered useful polymorphisms for genetic linkage analysis.

Polypeptide : A protein or part of a protein made of a chain of amino acids joined by a peptide bond.

Polyploid : A plant, or of a plant, with more than two sets (diploid) of the basic chromosome number (haploid).

Polysiphonous : Having many filaments or pericentral cells as in the red alga Polysiphonia.

Polysporangium : A sporangium producing many spores, homologous with a tetrasporangium.

Polystichous : Type of thallus constructed of true parenchyma as in some brown algal orders.

Polytypic : Of a taxon containing two or more tax of lower ranks.

Pome : It pertains to the characteristic fruit of the apple family such as apple, pear or quince, in which the edible flesh grows from the greatly tumefied receptacle and not from the carpels.

Population Genetics : The study of variation in genes among a group of individuals.

Poricidal : Opening by pores, like a poppy capsule.

Positional Cloning : A technique used to identify genes, usually those that are associated with diseases, based on their location on a chromosome.

Posterior : On the side next to the axis.

Potassium (K) : Chemical involved in regulation of protein synthesis and starch production in plants. Plants short in potassium may show stunted growth, weak stems and root systems, and/or spotted and curled leaves.

Pour Plate : The method of performing a plate count of microorganisms.

P-protein : This is a fibril protein which is responsible for plugging sieve pores and precludes outflow if sieve elements are damaged.

P-protein Plug : It relates to the obstruction of a sieve region or sieve plate by bast protein.

Predator : Organism that feed off or prey on other organisms for survival.

Premature Chromosome Condensation (PCC) : A method of studying chromosomes in the interphase stage of the cell cycle.

Prickle : A hard, pointed outgrowth from the surface of a plant, involving several layers of cells, but not containing a vascular system.

Primary Growth : A type of growth whereby a plant grows in length. Examples include the growth of roots and shoots.

Primary Pit Field : It is the region of the primary cell wall which is particularly thin and consists of many plasmodesmata.

Primary Producer : Any green plant which has the ability to convert light energy or chemical energy into organic substance.

Primary Tissue : Any tissue which is directly derived from distinction of an apical meristem or leaf primordium.

Primer : Short preexisting polynucleotide chain to which new deoxyribonucleotides can be added by DNA polymerase.

Primocane : The first-year (usually flowerless) cane or shoots of Rubus.

Privacy : In genetics, the right of people to restrict access to their genetic information.

Probe : Single-stranded DNA or RNA molecules of specific base sequence, labelled either radioactively or immunologically, that are used to detect the complementary base sequence by hybridisation.

Proboscis : An elongated mouth organ which is an important feeding appendage in organisms.

Procambium : A primary meristem of roots and shoots that forms the vascular tissue.

Procarp : Carpogonial branch adjacent to one or more auxiliary cells in Rhodophyta.

Processes : Outgrowth or projections from a surface.

Prochlorophytes : A class of procaryotes that possess both chlorophyll A and B, and is considered to be nearly associated to the antecedents of plastids in algae and plants.

Procumbent : Having stems trailing or spreading over the ground.

Producer : A photosynthetic green plant or chemosynthetic bacterium, that comprises of the first trophic level in a food chain.

Proembryo : This term relates to the cells which are forced into the endosperm and afterwards become the embryo, in seed bearing plant's embryos.

Prokaryotes : Organisms which do not possess true nucleus or membrane-bounded cell organelles such as eubacteria, cyanobacteria, and archaebacteria.

Proliferous : Plants which bear adventitious buds on the leaves or flowers, such buds being capable of rooting and forming separate plants. e.g. bulbils, epiphyllous plantlets.

Prominent : Standing out beyond some other part (dimin. prominulous).

Promoter : A DNA site to which RNA polymerase will bind and initiate transcription.

Promoter Region : The area of a cistron in which control molecules and RNA polymerases bind during the process of cistron activation and transcription.

Pronucleus : The nucleus of a sperm or egg prior to fertilisation.

Prop Root : An adventitious root which holds the plant, as the aerial roots of the Rhizophora mangle tree or of maize.

Propagule : A structure with the capacity to give rise to a new plant, e.g. (1) a seed, (2) part of the vegetative body capable of independent growth if detached from the plant.

Prophase : The initial phrase of mitosis or meiosis in eukaryotic cellular division, during which the nuclear envelope breaks down and filaments of chromatin form into chromosomes.

Proplastid : A cytoplasmic cell organelle from which a plastid originates and develops.

Protandrous : Describing a plant in which the release of pollen precedes and does not overlap the period of stigma receptivity.

Protein : A large molecule composed of one or more chains of amino acids in a specific order; the order is determined by the base sequence of nucleotides in the gene that codes for the protein. Proteins are required for the structure, function, and regulation of the body's cells, tissues, and organs; and each protein has unique functions. Examples are hormones, enzymes, and antibodies.

Protein Sequencing : This is a process that includes determining the amino acid sequences of its constituent peptides; and also finding out what compliance it follows and if it is complexed with any non-peptide molecules.

Proteome : Proteins expressed by a cell or organ at a particular time and under specific conditions.

Proteomics : The study of the full set of proteins encoded by a genome.

Prothallus : A tiny, flat, and gentle structure developed by a spudding spore, that has sex organs, and is the gametophyte of ferns and some other plants. Its structure resembles a leaf.

Proto : Prefix meaning 'first'.

Protoderm : A thin outer layer of the meristem in embryos and growing points of roots and stems, which gives rise to the epidermis.

Protogynous : Describing a plant in which stigma receptivity precedes and does not overlap the period of pollen release.

Protologue : The original or first publication of a taxonomic name.

Protonema : A threadlike structure created by sprouting of the spores in small leafy-stemmed flowerless plants and other related plants, and from which the leafy plant, that has the sexual organs, develops as a sidelong or terminal branch.

Protoplast : The living substance of a plant, including the protoplasm and cytomembrane after the cell wall has been removed.

Protoplast Fusion : A method by which two energids are coalesced to create hybrid cells that can develop into mature hybrid organisms; normally performed on plants.

Protostele : When a plant's vascular tissue develops in a solid central bundle, it is said to have a protostele.

Protuberance : A swelling or bump on the surface.

Proximal : Nearest the axis or base.

Pruinose : Frosted; covered with a waxy powdery secretion on the surface; having a "bloom".

Psuedo : Prefix meaning 'false'.

Pseudodichotomous : Apparently dichotomous with a dormant terminal bud and two equal lateral branches.

Pseudoelaters : Moisture-sensitive cells produced in the sporangium of hornworts.

Pseudogene : A sequence of DNA similar to a gene but nonfunctional; probably the remnant of a once-functional gene that accumulated mutations.

Pseudoparenchyma : Tissue resembling parenchyma with large, thin-walled cells, but developmentally filamentous.

Pseudopinnate : Term describing a lateral shoot resembling a pinnate leaf, i.e. a shoot with limited growth and with simple leaves arranged in 2 rows like the leaflets of a pinnate leaf.

Pseudowhorled : Of leaves, arranged in clusters on the stem, the clusters separated by regular intervals, usually produced behind a scaly bud.

Psychrotroph : An organism that is able to grow at zero degrees and above twenty degrees Celsius.

Pteridophyte : Plant in which the sporophyte generation is the larger phase and in which the gametophyte lives an existence independent of its parent sporophyte. Pteridophytes are almost all vascular plants, and include the lycophytes, trimerophytes, sphenophytes, and ferns.

Pteridosperm : An extinct group of seed plants which bore fernlike leaves.

Puberulent : Minutely pubescent, the hairs soft and very short, scarcely visible to the naked eye.

Puberulous : A dense covering of very short soft hairs, minutely pubescent.

Pubescent : Hairy or downy; covered with short, soft hairs.

Pulmonate : Land Snails and other air breathers belonging to Pulmonata Subclass and Sorbeconcha Clade.

Pulverulent : Dusty or chalky, as applied to the powdery coating on the stems and leaves of some plants.

Pulvinate : Cushion- or mat-like.

Pulvinule : The pulvinus at the base of a petiolule.

Pulvinus : The swelling at the base of the petiole, often capable of changing form to bring about movement of leaf, sometimes glandular or responsive to touch. A similar swelling near the apex of a petiole is referred to as an upper pulvinus. pl. pulvini.

Punctate : Dotted with pits or with translucent, sunken glands, or with coloured dots.

Punctiform : Reduced to a mere dot or point.

Punctum : A dot or pit; hence punctate, covered with dots or pits (pl. punctae).

Pungent : (1) ending in a stiff, sharp point; (2) having an acrid taste or smell.

Pure Culture : A microorganism population of a single strain.

Purebred Line : The homozygous dominant and homozygous recessive genetic constitution of a line which are selfed and utilised in spawning experiments.

Purine : A nitrogen-containing, double-ring, basic compound that occurs in nucleic acids. The purines in DNA and RNA are adenine and guanine.

Pustule : A low projection like a blister or pimple, larger than a papilla; hence pustular, pustulate.

Putative : Reputed, generally regarded as such, supposed; eg. putative hybrids.

Pycnoxylic : Wood in which there is little or no parenchyma tissue among the xylem is called pycnoxylic. Conifers and flowering plants have pycnoxylic wood.

Pyramidal : Pyramid-shaped, broadest at or near the base.

Pyrene : The endocarp and enclosed seed of a drupaceous fruit.

Pyrenoid : A proteinaceous structure that is found within the chloroplast of specific algae and nonvascular plants, which is believed to be related to starch deposition.

Pyrethrins : A class of chemicals found in many insecticides. They are a contact and a stomach poison which are lethal to most insects. They break down quickly in sunlight. Derived from the flowers of *Tanacetum cinerarifolium* (pyrethrum-related to the tansy and feverfew.).

Pyrimidine : A nitrogen-containing, single-ring, basic compound that occurs in nucleic acids. The pyrimidines in DNA are cytosine and thymine; in RNA, cytosine and uracil.

Q

Quadri : Prefix meaning 'four'.

Quadrifarious : Arranged in four close-set rows along the stem.

Quantitative Trait : These traits are controlled by various genes and environmental factors. They are measured on a continuous scale.

Quiescence : Every plant requires some specific environmental conditions for its proper functioning and rapid growth. The growth or germination of the seeds or plants are hampered if these environmental conditions are not satisfied. This is termed as 'quiescence'.

Quiescent Centre : Quiescent centre is the portion of the root situated at the apex of the plant tissue i.e. meristem in which cell division does not occur.

Quinate : With five nearly similar structures from a common point.

Quinque : Prefix meaning five.

R

R Replum : A longitudinal partition in fruits of the family Brassicaceae.

Raceme : A flower cluster with an elongated main axis but with flowers borne on pedicels of equal length.

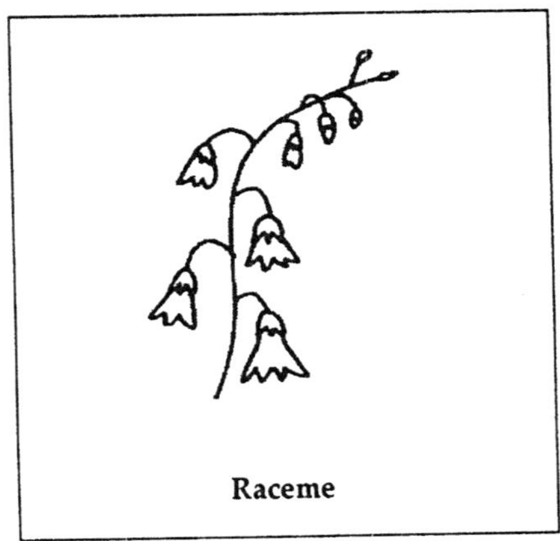

Raceme

Raceme-like : Applied to a simple inflorescence ending in a floral bud in which the flowers are stalked, i.e. resembling a raceme but determinate; also used for conflorescences of similar form, particularly where there has been reduction from more complex types.

Racemose : Raceme-like or bearing racemes.

Rachilla : The axis of a grass spikelet above the glumes, also the axis in sedge spikelets.

Rachis : Rachis is the extension of the axis of petiole or leafstalk in the compound leaf. All leaflets are attached to the rachis.

Radial, Raiate : Spreading as in the spokes of a wheel, eg. radially symmetric.

Radiate : Describing a flower head in the *Asteraceae* that contains both ray and disk flowers.

Radiation : Evolution of multiple species from a single ancestry, but these species have morphological differences, however, they coexist in the same habitat or spread to different habitats or they have a change of ecological role.

Radiation Hybrid : A hybrid cell containing small fragments of irradiated human chromosomes. Maps of irradiation sites on chromosomes for the human, rat, mouse, and other genomes provide important markers, allowing the construction of very precise STS maps indispensable to studying multifactorial diseases.

Radical : Of leaves arising at the base of the stem and forming a rosette or tuft.

Radicant : Rooting, usually applied to stems and leaves.

Radicle : Extension of the axis of petiole or leafstalk in the compound leaf. All leaflets are attached to the rachis.

Radioimmunoassay : An immunological assay that makes use of radioactive antibodies or antigens to detect certain substances.

Radiospermic : Having seeds which are round or ovoid.

Rainforest (Closed Forest) : A forest dominated by broad-leaved trees with dense crowns that form a continuous layer (canopy) and with one or more of the following growth forms.

Ramiflory : The production of flowers and fruits behind the current foliage on woody branches formed in previous, but recent, seasons. adj. ramiflorous.

Range : A particular geographical area in which particular species of organisms are found.

Rank : (1) a vertical row; (2) leaves that are 2-ranked are in 2 vertical rows, and may be alternate or opposite.

Raphe : Elongate opening in the valves of some pennate diatoms.

Rare : A specie of an organism found in very small numbers and hence, visible with lot a of effort only for a short duration.

Rattle : Shed skin, which is often seen on tail of a rattlesnake, used to make a rattling sound in order to deter predators.

Ray : Series of parenchyma cells that are radially arranged along the vascular region of the xylem and the phloem. These parenchyma cells transport food, water and other materials laterally in the roots and stems of woody plants.

Ray Floret (RayFlower) : A zygomorphic flower in many Asteraceae, usually formed towards the periphery of the head and with the corolla extended into a strap-shaped ligule.

Re : Prefix meaning 'backwards'.

Reaction Centre : A photosynthetic complex containing chlorophyll and other compounds.

Reaction Wood : Is formed when a woody plant encounters mechanical stress, as caused by wind exposure, soil movement and excess snow fall.

Reannealing : The process seen on cooling, where two complementary strands of DNA hybridise back into a single strand.

Recalcitrant : Resistance of an organism to a microbial attack.

Receptacle : The expanded apex of a flower stalk which bears the floral organs, either such structures as individual petals, sepals etc., or entire flowers in head-like inflorescences such as is typical of the *Asteraceae*.

Recessive Gene : A gene which will be expressed only if there are 2 identical copies or, for a male, if one copy is present on the X chromosome.

Recessive Trait : It is the trait that reflects in the phenotype only when the dominant gene is absent. Eg : Colour blindness.

Reciprocal Translocation : When a pair of chromosomes exchange exactly the same length and area of DNA. Results in a shuffling of genes.

Reclinate : Reclining, turned or bent downwards upon some other part.

Recombinant Clone : Clone containing recombinant DNA molecules.

Recombinant DNA : DNA molecule created either by crossing over in meiosis or under laboratory environement (in vitro). It is formed when DNA from atleast two organisms is taken.

Recombinant DNA Molecules : A combination of DNA molecules of different origin that are joined using recombinant DNA technologies.

Recombinant DNA Technology : Procedure used to join together DNA segments in a cell-free system (an environment outside a cell or organism). Under appropriate conditions, a recombinant DNA molecule can enter a cell and replicate there, either autonomously or after it has become integrated into a cellular chromosome.

Recombination : Process by which genetic elements in two separate genomes are brought together in one unit. This is an important step in gene therapy.

Recumbent : Leaning or reposing upon the ground.

Recurrent : Running or proceeding backwards towards the axis or costa.

Recurved : Curved backwards (and hence usually downwards), e.g. of the margins of a leaf.

Red Tide : Marine phenomenon in which a reddish tint is formed on the water due to the sudden growth of cells in certain protozoa or red algae.

Reflexed : Abruptly bent or curved downward.

Regular : Describes a flower with petals or sepals all of equal size and shape, i.e. radially symmetrical or capable of being divided into mirror images on either side of any plane that passes through the centre.

Regulatory Region or Sequence : A DNA base sequence that controls gene expression.

Reniform : Kidney-shaped or rounded with a notch at the base.

Repand : With an undulating margin, less strongly wavy than 'sinuate'

Repetitive DNA : Sequences of varying lengths that occur in multiple copies in the genome; it represents much of the human genome.

Replication : Conversion of one double stranded DNA molecule into two identical double stranded DNA molecules.

Repression : Process by which an enzyme synthesis is suppressed due to the presence of certain external substance.

Reproduction : It is the birth of a new organism born either by sexual or asexual means.

Resin Canal : Tubular duct present in coniferous trees and seeds, which is lined with resin secreting cells.

Resin Duct : Long narrow channel filled with resin.

Resolution : Degree of molecular detail on a physical map of DNA, ranging from low to high.

Respiration : Cellular breakdown of sugar and other food molecules, in which a part of energy is utilised. If oxygen is utilised during the breakdown, it is known as aerobic respiration or else it is termed as anaerobic respiration.

Restriction Enzyme, Endonuclease : A protein that recognises specific, short nucleotide sequences and cuts DNA at those sites. Bacteria contain over 400 such enzymes that recognise and cut more than 100 different DNA sequences.

Restriction Fragment Length Polymorphism (RFLP) : Variation between individuals in DNA fragment sizes cut by specific restriction enzymes; polymorphic sequences that result in RFLPs are used as markers on both physical maps and genetic linkage maps. RFLPs usually are caused by mutation at a cutting site.

Restriction-enzyme Cutting Site : A specific nucleotide sequence of DNA at which a particular restriction enzyme cuts the DNA. Some sites occur frequently in DNA (e.g., every several hundred base pairs); others much less frequently (rare-cutter; e.g., every 10,000 base pairs).

Resupinate : Twisted through 1800 as in the ovary of most Orchidaceae.

Reticulate : Forming a network or reticulum; e.g. of veins.

Reticulate Venation : Reticulate venation is a thin, flat, laminar like structure of a leaf, featuring a net-like pattern of the veins, structured for the purpose of photosynthesis.

Reticulated : Species whose veins or nerves are like threads of a net, arranged in a network.

Reticulum : A network (of veins or other linear structures), formed by repeated branching and anastomosis (viz.).

Retinaculum : (1) a hooklike structure to which another structure is tethered, as in Orchidaceae and Asclepiadaceae (the structure to which pollen masses are attached) or in Acanthaceae (the persistent stalk of an ovule); (2) the marginal outgrowth from a spadix, as in Zosteraceae. pl. retinacula.

Retrorse : Bent backward or downward, reflexed.

Retroviral Infection : The presence of retroviral vectors, such as some viruses, which use their recombinant DNA to insert their genetic material into the chromosomes of the host's cells. The virus is then propogated by the host cell.

Retrovirus : Common type of plant virus whose genetic material is single-stranded RNA.

Retuse : Having the apex rounded and with a small notch.

Reverse Migration : A phenomena where the migrating organism migrates in the opposite direction, normal to other migrating species.

Reverse Transcriptase : An enzyme used by retroviruses to form a complementary DNA sequence (cDNA) from their RNA.

The resulting DNA is then inserted into the chromosome of the host cell.

Reverse Transcription : Process of copying information from RNA to DNA.

Revolute : Having the margins inrolled toward the underside.

Rheophyte : A flood persistent plant, living between the high and low water levels of rivers.

Rhizobacteria : Bacteria that are found in roots, where they aggressively colonise.

Rhizobia : Bacteria capable of living symbiotically in leguminous plant roots, from where they receive energy and commonly fix molecular dinitrogen.

Rhizoid : n. A cellular outgrowth of a plant that usually aids in anchoring to the surface and increasing surface area to acquire water or nutrients; found in mosses, liverworts, and hornworts.

Rhizome : A horizontal underground stem, such as found in many ferns, where only the leaves may stick up into the air; sphenophytes (horsetails and their relatives) spread via rhizomes, but also produce erect stems.

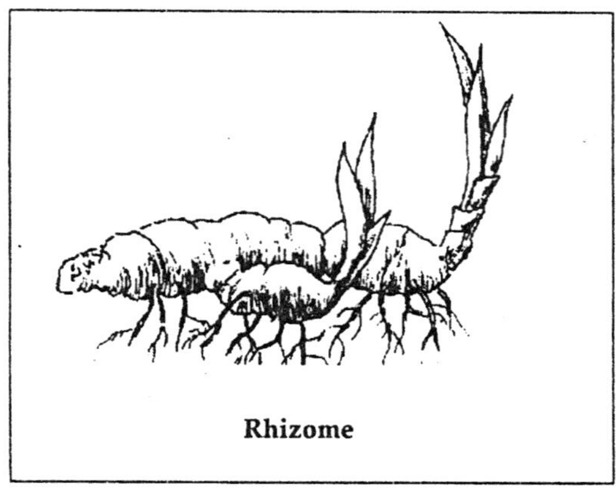

Rhizome

Rhizomorph : Mass of fungal hyphae that are organised in long, thick strands with a darkly pigmented outer rind that contains specialised tissues for absorption and water transport.

Rhizophore : A specialised leafless stem which bears roots, as in Selaginella.

Rhizoplane : Plant root surface and strongly adhering soil particles.

Rhizosphere : The zone of soil immediately adjacent to plant roots in which the activity and type of microorganisms present differ from that in the rest of the soil.

Rhizosphere Competence : Ability of an organism to colonise the rhizosphere.

Rhombic : Having the form of a 2-dimensional diamond-shaped figure.

Rhomboidal : Approaching the shape of a rhombus, i.e., with the outline of an equilateral parallelogram with oblique angles. In a rhomboid only opposite sides are equal.

Ribonucleic Acid (RNA) : Type of molecule containing large amount of nucleotide units, wherein each nucleotide contains three elements- nitrogenous base, a ribose sugar, and a phosphate. It is involved in protein synthesis.

Ribose : The five-carbon sugar that serves as a component of RNA.

Ribosomal RNA (rRNA) : A class of RNA found in the ribosomes of cells.

Ribosome : Cell organelle composed of proteins and ribonucleic acid (RNA), which is responsible for protein synthesis.

Ribosomes : Small cellular components composed of specialised ribosomal RNA and protein; site of protein synthesis.

Riparian : Of plants growing by rivers or streams.

RNA (Ribonucleic Acid) : A chemical found in the nucleus and cytoplasm of cells; it plays an important role in protein synthesis and other chemical activities of the cell. The structure of RNA is similar to that of DNA. There are several classes of RNA molecules, including messenger RNA, transfer RNA, ribosomal RNA, and other small RNAs, each serving a different purpose.

Root : Organ of the plant situated below the ground and absorbs water and mineral salts. Buds, leaves or nodes are absent in root.

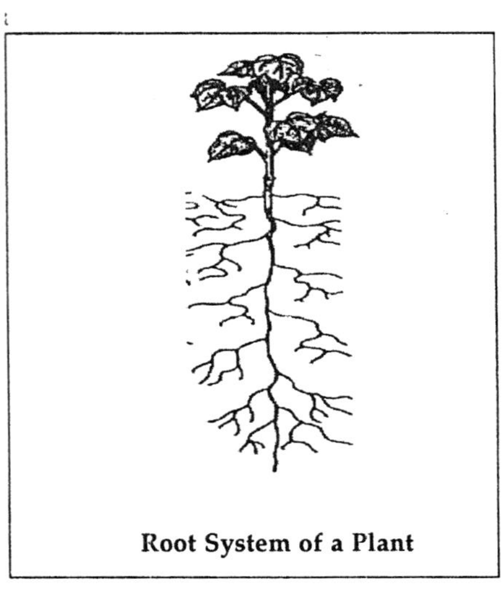

Root System of a Plant

Root Cap : The tip of a root containing cells which protect the growing meristem. The root cap also secretes a substance that acts as lubrication so the growing root tip can manoeuvre through the soil.

Root Hair : Hairlike outgrowth arising through the epidermal cell of the root. Located just behind the tip of the root, this root hair helps absorb water and nutrients from the soil.

Root Nodule : Plants forming symbiotic associations with nitrogen-fixing bacteria, exhibit swelling in their roots, at the region where the bacteria comes in contact with the root.

Root Pressure : Upward water pressure in roots where water is pushed up the stele by the pumping of minerals into the xylem.

Rootstock : A swollen region at the junction of root system and stem, mostly below ground level.

Rosette : A group of organs radiating from the centre, especially with numerous overlapping leaves appressed to the soil; hence rosulate.

Rostellum **:** In orchids, a projection of the upper edge of the stigma in front of the anthers.

Rostral Scale **:** Is a scale present on the tip of the upper jaw of a snout, usually seen in snakes.

Rostrate **:** Having a beak or beaklike form.

Rotate **:** A rotate corolla is wheel-shaped with a short tube and a wide horizontally flaring limb.

Rotound, Rotundate **:** Rounded, almost circular.

Rounded **:** Smallest size elliptical, spherical egg.

Rrosette **:** A series of whorls of leaves or leaflike structure produced at the base of the stem, just above the ground.

Rubescent **:** Becoming red or reddish.

Ruderal **:** Growing in disturbed habitats, weedy.

Rudiment **:** An imperfectly developed organ, a vestige.

Rudimentary **:** Arrested at an early stage of development.

Rugose (Wrinkled) **:** Covered with coarse lines or furrows.

Ruminate **:** (1) of a surface or tissue, with an irregular, involuted outline, as in a rumen; (2) mottled in appearance.

Runcinate **:** Sharply incised or pinnatifid with the segments facing backwards.

Runner **:** Slender creeping stem that contains long inter-nodes, growing horizontally along the surface of the ground. Eg: Strawberry plant.

Rupestral **:** Growing among rocks or on rock walls.

S

Sagenoid : Of anastomosing venation with regular areoles with included, free, often branched veinlets pointing in all directions; like species of Tecaria.

Sagittate : Shaped like an arrowhead, with the two lobes at the base acute and retrorse; e.g of a lamina or sometimes applied to the base of a lamina.

Saline Soil : A subset of salty soil which contains large quantities of soluble salt compounds such as sodium, calcium, chloride, etc.

Salty Soil : Soil which has high concentrations of salt. Found in coastal and arid regions. The abudance of salt in the soil causes water loss from plants via the root system have a detrimental effect on the plants ability to uptake mositure and nutrients into the plant system. Subsets of salty soils are saline soils and sodic soil.

Salverform : With a slender tube abruptly expanded into a rotate limb.

Samara : An indehiscent winged fruit with a single seed as in maple, elm or ash.

Sandy Soil : A type of soil with excellent drainage due to its characteristic large grains of sand.

Sanitisation : Elimination of pathogenic or harmful organisms, including insect larvae, intestinal parasites and weed seeds.

Saprobe : Saprobes are heterotrophs which contribute to the various nutrient cycles by feeding on decomposing organic matter.

Saprophyte : An organism using decaying or non-living organic matter for nourishment; hence saprophytic.

Saprophytic : Deriving food from dead or decaying organic material in the soil and usually lacking in chlorophyll.

Sapwood : Sapwood is the outer wood which carries water from roots to the leaves in order to facilitate water storage for future use.

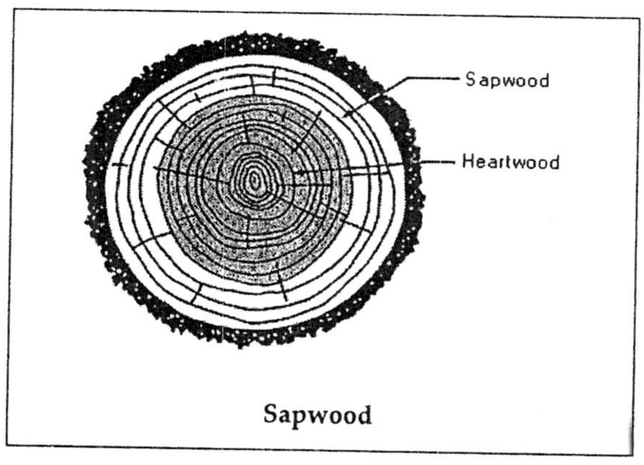

Sapwood

Sarmentose : Producing long, flexuose runners or stolons.

Satellite : A chromosomal segment that branches off from the rest of the chromosome but is still connected by a thin filament or stalk.

Satiny : Of an indumentum of fine haris closely apprese to form a complete, very smooth, more or less shiny cover.

Savannah : Tropical grassland biome characterised by scrubland, trees, grazing mammals, and a seasonal cycle of rainy and dry seasons.

Saw : A tool with sharp teeth used for pruning trees and shrubs.

Saw-Scaling : Action of a snake curving its body in concentric curves and rasping its keeled scales together to make a sawing sound as a warning.

Saxatile : Growing among rocks or in rocky, arid situations.

Saxitoxin : Neurotoxin produced by dinoflagellate *Gonyaulax*.

Scabrid, Scarbrous : Rough to touch due to minute, hard projections (dimin. scabridulous, scaberlous).

Scabrous : Rough to the touch; having the surface rough with minute hard processes or very short rigid hairs.

Scaffold : In genomic mapping, a series of contigs that are in the right order but not necessarily connected in one continuous stretch of sequence.

Scalariform : Ladder-like, the markings suggestive of a ladder, eg. venation.

Scale : (1) any thin and often scarious body, often a reduced or rudimentary leaf, e.g. covering a dormant bud; (2) a thin flap of tissue, e.g. at the base of stamens; (3) a small papery surface structure on stems and leaves.

Scandent : Climbing, usually applied in cases where special climbing organs are not developed.

Scape : A tiny stem like first segment in an insect's antennae, as the shaft of a feather.

Scarify : To roughen, score or scrape the hard, outer coating of a seed to assist in the absorption of moisture before germination, a process that many desert wash seeds require.

Scarious : Thin, dry, membranous and more or less translucent.

Schizocarp : A dry fruit which splits into individual carpels, each of which is called a mericarp or coccus.

Schizocarpic Capsule : A schizocarp in which the individual cocci or mericarps dehisce, as in some Rutaceae.

Scion : A shoot or sucker growing from a mature plant, generally used for grafting purpose.

Sclereid : An irregular sclerenchyma cell found in the parenchyma of some plants, in nutshells, and seed coats.

Sclerenchyma Cell : A plant cell that provides rigid support. Characteristic thick secondary cell walls that contain lignin, a strengthener at maturity. There are two types of sclerenchyma cells.

Scleromorphic : Hard and with a large amount of fibrous tissue.

Sclerophyll : A plant with hard, stiff leaves. adj. sclerophyllous.

Sclerophyllous : With stiff, firm leaves.

Sclerotium : Modified fungal hyphae that form a compact and hard vegetative resting structure with a thick pigmented outer rind.

Scorpioid : Describing a coiled inflorescence.

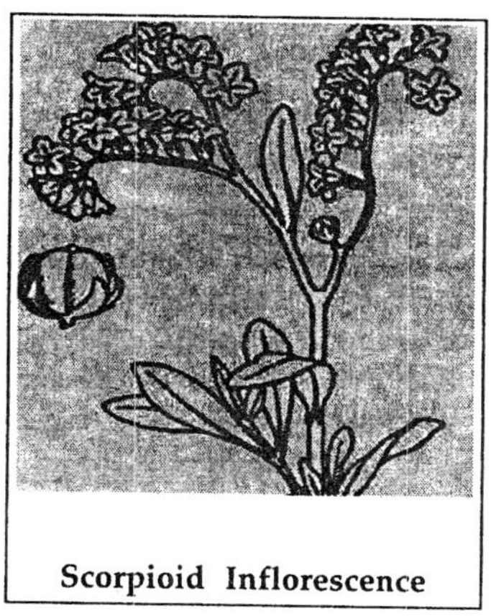

Scorpioid Inflorescence

Scramblers : A type of vine that does not have any adapted features to help secure itself. This type of vine needs to be staked or tied to a support.

Scribbles : Irregular lines on the barks of some eucalypts, being the old tunnels burrowed by moth larvae between the bark layers and exposed when the outer layer sheds.

Scrobiculate : Marked with minute depressions.

Scrub : A community dominated by shrubs.

Scurfy : Covered with small scale-like or brainlike particles or projections.

Scute : A large well defined dermal bony plate or horny plate found on many reptiles.

Scutellum : A shield; hence scutate, scutelliform, shield-shaped (pl. scutella).

Secondary Growth : Growth in a plant which does not occur at the tips of the stems or roots. Secondary growth produces wood and bark in seed plants.

Secondary Metabolite : Product of intermediary metabolism released from a cell, for example, antibiotic.

Secondary Phloem : Secondary phloem is the phloem which is derived from vascular cambium.

Secondary Thickening : The production of additional vascular and supporting tissue through the activities if a vascular cambium.

Secondary Tissues : Secondary tissues are the tissues of the secondary plant body which are produced by vascular cambium.

Secondary Xylem : Secondary xylem is the xylem that is derived from vascular cambium.

Sect : Suffix meaning 'deeply divided or lobed almost to the base' eg. pinnatisect, palmatisect.

Section : A subgroup of a genus used to identify closely related species.

Secund : Flowers or other structures arranged on, or turned to one side of an axis, e.g. inflorescence of many Grevillea species.

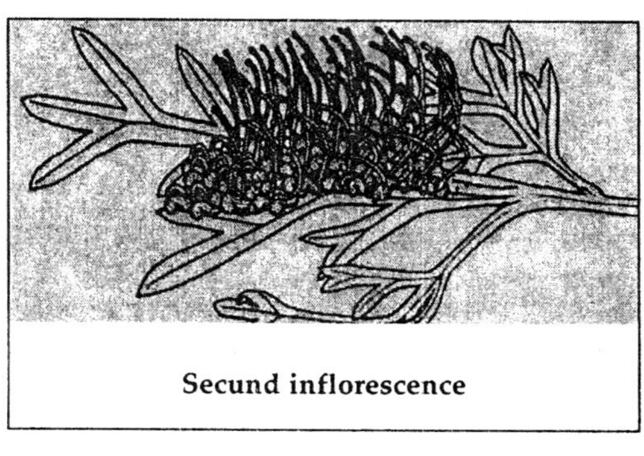

Secund inflorescence

Sedentary : Organism that are nonmigratory in nature, which means they move about little or not at all from their habitats.

Seed : An adapted feature of terrestial plants in which an embryo with food is stored and a protected casing.

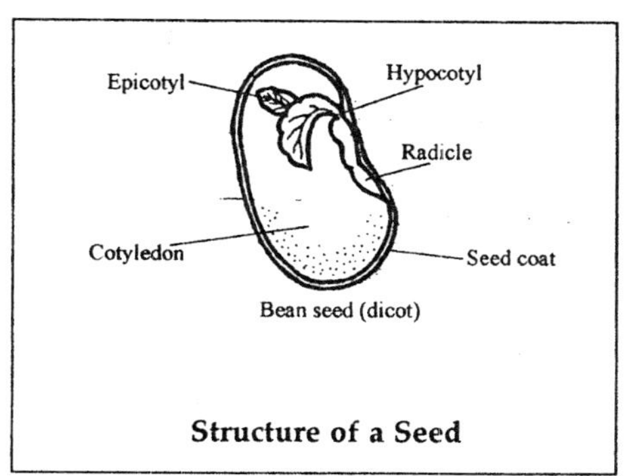

Structure of a Seed

Seed Coat : Seed coat, also referred to as testa, is the protective outer covering of seeds of various flowering plants.

Segment : A free or almost free part or subdivision of an organ. Calyx and corolla segments called sepals and petals respectively. Undifferentiated segments are called tepals.

Segregation : The normal biological process whereby the two pieces of a chromosome pair are separated during meiosis and randomly distributed to the germ cells.

Selective Medium : A medium that is biased in allowing only certain types of microorganisms to grow.

Selectively Impermeable Membrane : It is a barrier which regulates the movement of substances, allowing some substances to pass rapidly, and others to slow down.

Selectively Permeable Membrane : It is a membrane that facilitates the transmission of certain molecules through it by the process of diffusion.

Self Seed : Referring to a plant that regenerates itself by dropping its own seeds. Characteristics of annuals. Synonym of self sow.

Selfing : Selfing is a process wherein a plant's stigma is pollinated with pollen either from the same plant, or from a plant of identical genetic constitution.

Semi : Prefix meaning half.

Semi-double Flower : A type of flower which has double the normal amount of petals arranged in two-three rows of petals.

Semi-Precocial : This term refers to hatchlings that are capable of leaving the nest but are dependent on their parents for their feed.

Sepal : The sepal is the outermost part of a flower, resembling a leaf, which forms the calyx of the flower and surrounds its reproductive organs.

Sepaloid : Resembling a sepal, i.e. not petaloid.

Septate : Divided by internal transverse partitions.

Septicidal : Said of a capsule, longitudinally dehiscent through the ovary wall at or near the centre of each septa, preserving each locule as an intact entity.

Septicidal Capsule : Dehiscing by splits along the sutures between adjacent carpels.

Septum : A partition or cross-hall; hence septate, internally divided by transverse partitions.

Sequence Assembly : A process whereby the order of multiple sequenced DNA fragments is determined.

Sequence Tagged Site (STS) : Short (200 to 500 base pairs) DNA sequence that has a single occurrence in the human genome and whose location and base sequence are known. Detectable by polymerase chain reaction, STSs are useful for localising and orienting the mapping and sequence data reported from many different laboratories and serve as landmarks on the developing physical map of the human genome. Expressed sequence tags (ESTs) are STSs derived from cDNAs.

Sequencing : Determination of the order of nucleotides (base sequences) in a DNA or RNA molecule or the order of amino acids in a protein.

Sequencing Technology : The instrumentation and procedures used to determine the order of nucleotides in DNA.

Serial Dilution : Series of stepwise dilutions, normally done in sterile water, which is done to reduce microorganism populations to manageable numbers.

Seriate : Suffix indicating arranged in rows, eg. uniseriate, biseriate, multiseriate etc.

Serpentine : Refers to soils that are low in calcium and high in magnesium and iron, derived from greenish or gray-green rocks that are essentially magnesium silicate, other characteristics of which are a high nickel and chromium content, and a low content of nutrients such as nitrogen.

Serrate : A margin armed with sharp teeth which point forward like the teeth of a saw.

Serrulate : Serrate with very small teeth.

Sessile : The term sessile, meaning without a stalk, is most often used in context of plants whose flowers or leaves grow directly from the stem.

Seta : Seta is a botanical term used to refer to the stalk of the capsule, which is located in between the foot and the sporangium.

Sex Chromosome : The X or Y chromosome in human beings that determines the sex of an individual. Females have two X chromosomes in diploid cells; males have an X and a Y chromosome. The sex chromosomes comprise the 23rd chromosome pair in a karyotype.

Sex-linked : Traits or diseases associated with the X or Y chromosome; generally seen in males.

Sexual : Concerned with reproduction through the union of male and female gametes to form a zygote which develops into a new plant.

Shearing : A type of pruning in which random cuts are taken to achieve an even surface. This is used when maintaining hedges or topiaries.

Shears : Large clippers used to cut thick branches or stems during pruning.

Sheath : Tubular structure that is found either around a chain of cells or around a bundle of filaments.

Sheathing : Clasping or surrounding the stem.

Shell : A hard outer covering of an organism made up of carapace and plastron.

Shoot System : The above ground portion of a plant comprised of the stem/trunk, branches, flowers, and leaves.

Short Day Plant : A plant that responds (by flowering) to photoperiods that are shorter (i.e. during winter).

Shotgun Method : Sequencing method that involves randomly sequenced cloned pieces of the genome, with no foreknowledge of where the piece originally came from. This can be contrastedd with "directed" strategies, in which pieces of DNA from known chromosomal locations are sequenced. Because there are advantages to both strategies, researchers use both random (or shotgun) and directed strategies in combination to sequence the human genome.

Shrub : A much-branched woody plant less than 8 m high and usually with many stems. Tall shrubs are mostly 28 m high; small shrubs 12 m high; subshrubs less than 1 m high.

Siderochromes : The compounds that are synthesised by the microorganisms themselves, which are responsible for iron uptake.

Siderophore : A metabolite that is formed by some microorganisms, that forms a strong coordination compound with iron.

Sieve Area : Field of pores through which translocation occurs as in brown algal order Laminariales.

Sieve Cell : Sieve cells are conducting cells of the secondary phloem, which have a narrow diameter and are more elongated in shape as compared to the sieve tube members.

Sieve Plate : Sieve plates are the pores, in the cell walls of the plant, which facilitate the movement of liquid matter.

Sieve Tube : Sieve tube is a tube formed by cells joined end-to-end in order to facilitate the flow of nutrients in flowering plants.

Sieve-tube Member : Cells in a chain that form sieve tubes in phloem.

Silicle : A fruit similar to a silique, but much shorter, not much longer than wide.

Silicoflagellate : Chrysophycean organism with silicious skeleton.

Silicula : A fruit like a siliqua but less than 3 times as long as broad.

Siliqua : A dry dehiscent fruit derived from a superior ovary of two carpels and with 2 parietal placentas connected by a false septum, usually at least 3 times as long as broad, as in family Brassicaceae.

Silky : Covered with fine soft more or less straight appressed hairs aligned in the same direction, with a lustrous sheen and satin-like to the touch.

Silt : A type of soil particle that is 1/500 in. thick and is found in loam soil.

Similifacial (isobilateral) : Having structurally similar upper and lower surfaces.

Simple : Of a single piece or series; (1) of a leaf, with lamina not divided into leaflets, (2) of an inflorescence, unbranched with the pedicels arising from the main axis.

Simple Cone : A simple cone is a cone featuring only one axis or bearing only sporophylls.

Simple Fruit : A fruit that is derived from a single ovary. For example a cherry.

Simple Leaf : A simple leaf is a single leaf blade sporting a bud at the base of the leaf-stem.

Sinewy : Fluted; with strongly developed fluting.

Single Flower : A flower that has a single row of petals which contains all of the petals for that blossom.

Single Nucleotide Polymorphism (SNP) : DNA sequence variations that occur when a single nucleotide (A, T, C, or G) in the genome sequence is altered.

Single-gene Disorder : Hereditary disorder caused by a mutant allele of a single gene (e.g., Duchenne muscular dystrophy, retinoblastoma, sickle cell disease).

Sinistrorse : Turned to the left or spirally arranged to the left.

Sink : Sink is a botanical term used to refer to any tissue which receives the material that is transported by the phloem.

Sinuate : Strongly or deeply wavy, usually referring to a leaf margin.

Sinus : The space or division, usually on a leaf, between two lobes or teeth.

Sinus Membrane : A hyaline flap of tissue occurring in the base of the sinus in some ferns.

Siphonestele : A cylindrical stele with a hollow parenchymatous centre; amphiphloic siphonostele (= solenostele) has phloem on both the inside and outside of the cylinder, an ectophloic siphonostele (=medullated protostele) has phloem on the outside only.

Siphonostele : Siphonostele is a type of stele, usually characterised by the formation of cylinder surrounding the central pith and possessing leaf gaps.

Slime Layer : A diffuse layer found immediately outside the cell wall in certain bacteria.

Slime Mold : Microorganisms that are eukaryotic and which lack cell walls.

Sodic Soil : A type of salty soil that has large quantities of insoluble sodium. This soil type has poor drainage and is highly alkaline.

Softwood : Softwood refers to any of the various varieties of trees, usually coniferous, sporting narrow, needle like leaves.

Soil : Soil is made up of many different organic materials, minerals, and other nutrients. There are many types of soils such as : sandy, loam, clay, alkaline, acidic, and salty.

Soil Soakers : Used in irrigation systems, a perforated tube that allows large amount of water to be distributed to plants.

Soil Solarisation : The process of covering soil with plastic to trap heat underneath and kill off weeds.

Solarisation : A technique to control the growth of pathogens, wherein a plastic sheet is used to cover moistened soil in hot climates, thereby trapping the incoming radiation.

Solitary : Borne singly, not grouped together, e.g. of flowers not grouped into an inflorescence.

Somatic Cell : Any cell in the body except gametes and their precursors.

Somatic Cell Gene Therapy : Incorporating new genetic material into cells for therapeutic purposes. The new genetic material cannot be passed to offspring.

Somatic Cell Genetic Mutation : A change in the genetic structure that is neither inherited nor passed to offspring. Also called acquired mutations.

Sori : Clusters of spore sacs on a fern frond (Singular: Sorus)

Sorophore : Specialised sporangia-bearing lobes of the leaf margin in species of Schizaeaceae.

Sorus : A group or arrangement of sporangia in ferns (pl. sori).

Southern Blotting : Transfer by absorption of DNA fragments separated in electrophoretic gels to membrane filters for detection of specific base sequences by radio-labelled complementary probes.

Sp : Abbreviation for 'species'

Spade : A tool used to prepare soil. Has a narrower, shorter and flatter blade than a shovel.

Spading Fork : A tool often used in clay soils, it helps to break up large dirt clods.

Spadix : A spicate inflorescence with a thickened, often succulent axis, the whole often being surrounded by a spathe.

Spathe : A large bract or pair of bracts subtending and usually partially enclosing an inflorescence.

Spatulate : Oblong, wide and rounded at the apex but gradually narrowed near the base.

Species : The fundamental unit of taxonomy, a group of plants of the same morphology, capable of interbreeding, sometimes with +/- distinct varieties; similar species are grouped into genera.

Specific Activity : Expressed as micromoles formed per unit time per milligram of protein, this is the amount of enzyme activity units per mass of protein.

Specific Epithet : The second part of a scientific name which identifies the species.

Spectral karyotype (SKY) : A graphic of all an organism's chromosomes, each labelled with a different colour. Useful for identifying chromosomal abnormalities.

Spermatophytes : Spermatophytes are the plants that reproduce by means of seeds, instead of spores.

Spermosphere : The area seen around a germinating seed, where there is increased microbiological activity.

Sphagnum Moss : A type of moss found in bogs and used commercially to line hanging baskets. It is also used to make up peat moss.

Spicule : A short, pointed, epidermal projection.

Spike : A simple inflorescence, terminating in a non-floral bud, in which the flowers are sessile, i.e. a type of indeterminate inflorescence.

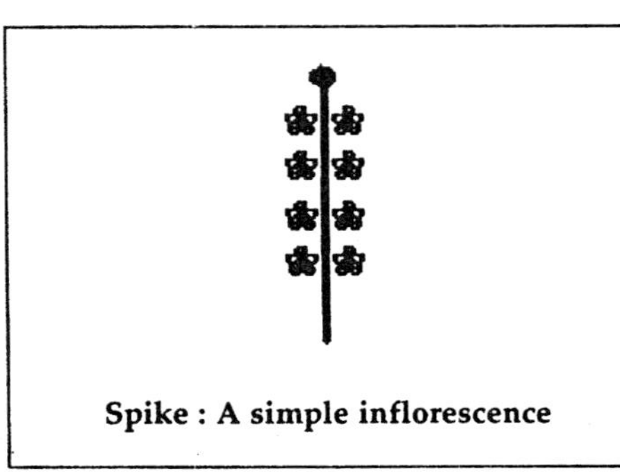

Spike : A simple inflorescence

Spikelet : The small partial inflorescence (unit) in Poaceae, Cyperaceae and Restionaceae, composed of an axis bearing glumes, most of which enclose a small flower.

Spike-like : Of a simple inflorescence, terminating in a floralbud and having the flowers sessile, i.e. a type of determinate inflorescence.

Spindle : Spindle is the underlying structure of microtubules which pulls away the chromosomes from the centre of the cell, towards the poles during the process of nuclear division.

Spindle-shaped : Broader in the middle, tapering at each end.

Spine : A stiff process with a sharp point, formed by a modification of a plant organ that contains vascular tissue, e.g. a lateral branch or a stipule.

Spinescent : Terminated by a spine; modified to form a spine.

Spinose : Having a stiff and tough acuminate tip.

Spinulose : Bearing very small spines.

Spiral : Of leaves or floral organs, borne singly at different levels on the axis, but not in a single vertical line; leaves borne spirally are said to be alternate on the stem.

Spirillum : The term spirillum is used to refer to the spirally twisted bacteria, resembling an elongated rod, which is usually found in stagnant water.

Splice Site : Location in the DNA sequence where RNA removes the noncoding areas to form a continuous gene transcript for translation into a protein.

Spongy : Having the texture of a sponge, the cells being separated by air spaces and containing air, as in pith or some seed coats.

Spongy Mesophyll : Spongy mesophyll are irregularly shaped and distinctly spaced parenchyma cells present in plant leaves.

Sporadic Cancer : Cancer that occurs randomly and is not inherited from parents. Caused by DNA changes in one cell that grows and divides, spreading throughout the body.

Sporangiophore : A stalk to which sporangia are attached.

Sporangium : The term sporangium is used to refer to a plant structure which produces and contains spores.

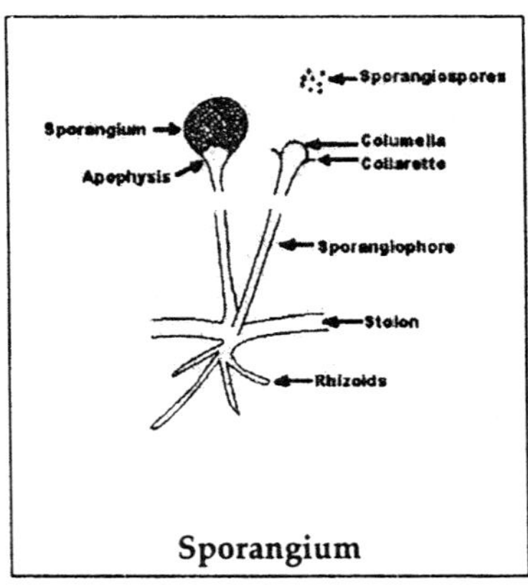

Sporangium

Spore : A spore is a unicellular asexual reproductive body in several nonflowering plants that facilitates the development of a new individual without sexual fusion.

Sporeling : A very young fern plant developing from the prothallas after fertilisation.

Sporocarp : A thick-walled body containing sporangia (eg. Marsilea).

Sporogenous : Of tissue or cells in which the spores are formed.

Sporophyll : Leaflike or foliaceous organ bearing reproductive parts or organs.

Sporophyte : The result of the mitotic division of the haploid spore. It can also form from the union of haploid gametes and later give rise meiotically to haploid spores.

Sport : A variation or mutation that occurs within the plant and differs from the rest of the plant.

Spp : Abbreviation for the plural of 'species'

Spread Plate : A technique for performing a plate count of microorganisms.

Sprinkler : A device used in irrigation systems to disperse water via a spray mechanism usually in a circular pattern.

Spur : A short, slowly-grown branchlet; a hollow sac-like or tubular extension of some part of a flower, usually producing nectar.

Spur Pruning : A way to prune grapevines in which canes that fruited are cut back, the spurs with buds that remain will fruit the following season.

Spy Hopping : A vertical rise out of the water or tall grasses performed by certain cetaceans or land mammals respectively.

Squamate : Having or producing scales.

Squamose : Scaly or scale-like (dimin. squamulose).

Squamule : A small papery scale; hence squamulose.

Squarrose : With spreading or divergent processes or scales.

Ssp : Abbreviation for 'Subspecies'

Stallion : A male horse which is more than four years old.

Stamen : The male organ of the flower consisting of the anther and the slender filament meant to hold it in position is known as the stamen.

Stamminate : Describing a male flower that contains one or more stamens but no functional pistils.

Staminate Flower : A flower with stamens but usually without pistils.

Staminode : A sterile stamen or other nonfunctional structure occupying the position and having the overall appearance of a stamen.

Standard : The tree rose with a round head atop a straight trunk is a classic example of a standard. A plant trained to look like a small tree.

Starch : A polysaccharide in plants used for the storage of glucose.

Start Codon : The term start condon is used to refer to a set of three nucleotides which indicate the initiation of information for the process of protein synthesis.

Stele : The central vascular tube in roots which includes the xylem and phloem.

Stellate : Starlike, with radiating branches and often referring to the pattern of hairs on the surface of a leaf.

Stem : The main axis or a branch of the main axial system of a plant, developed from the plumule of the embryo and typically bearing leaves and generally above the ground.

Stem Cell : Undifferentiated, primitive cells in the bone marrow that have the ability both to multiply and to differentiate into specific blood cells.

Sterile (barren) : (1) without reproductive structures, not producing seed, spores or pollen; (2) of seeds, spores or pollen, not capable of germination.

Sterile : Barren, not producing seed, pollen or spores capable of germination; of a plant lacking reproductive organs.

Sterilisation : The process whereby an object or surface is rendered free of any living microorganisms.

Stichidia : Specialised branch bearing tetrasporangia in some Rhodophyta, usually larger than vegetative branches.

Stigma : The sticky tip of a pistil. Or, the dense region of pigments found in many photosynthetic protists which is sensitive to light, and thus functions somewhat like a miniature eye.

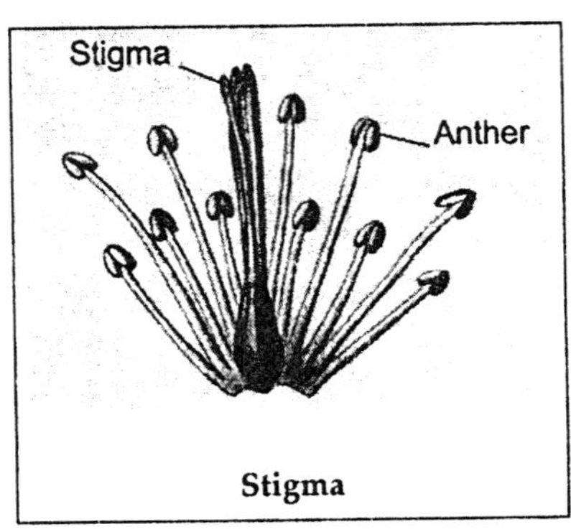

Stigma

Stilt Roots **:** Oblique adventitious roots from the stem, as in some mangroves and palms.

Stinging Hair **:** A hollow hair seated on a gland which secretes an irritating fluid, e.g. as in Dendrocnide species.

Stipate **:** Pressed together, crowed.

Stipe **:** (1) a stalk or support, e.g. of a gynoecium or carpel; (2) the petiole of a fern frond. adj. stipitate.

Stipel **:** A stipule-like organ at the base of a leaflet.

Stipitate **:** Borne on a stipe or stalk.

Stipular Spine **:** A spine formed from a stipule.

Stipule **:** One or a pair of appendages sometimes developed at the base of a leaf in many dicotyledons; can be leaflike, scarious or spinose.

Stipule Scar **:** The scar left by the fall of a stipule.

Stipules **:** Stipules are small leafy outgrowths, usually occurring in pairs, observed at the base of a leaf or the stalk.

Stolon **:** Modified stems that grow along the top of the soil, occasionally taking root and forming a new plant. Strawberry is an example.

Stoma **:** Stoma is a minuscule epidermal pore in the leaf or stem of the plant which allows gases and water vapour to pass through.

Stomata **:** Openings in the epidermis of a stem or leaf of a plant which permit gas exchange with the air. In general, all plants except liverworts have stomata in their sporophyte stage.

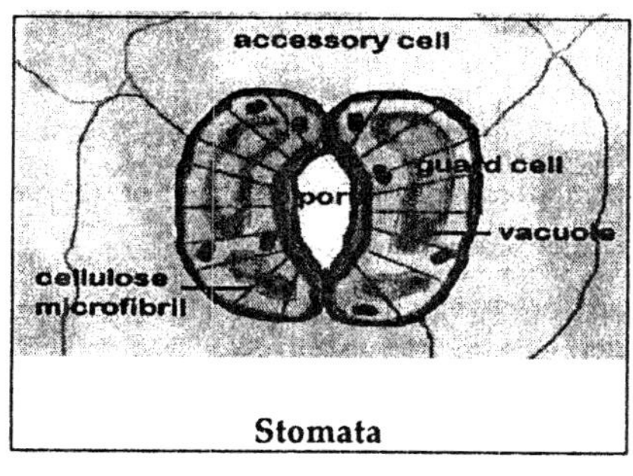

Stomata

Stomate : A small pore or opening on the surface of a leaf through which gaseous exchange takes place, i.e. the diffusion of carbon dioxide, oxygen and water vapour.

Stomium : The opening of the annulus in a sporangium through dehiscence occurs and the spores are shed, often marked by thin-walled, enlarged cells.

Stone : The hard, woody endocarp enclosing the seed of a drupe.

Stoop : To swoop down while in flight for catching a prey.

Stop Codon : The term stop codon is used to refer to the set of three nucleotides which indicate the termination of information for the process of protein synthesis.

Storage Polysaccharide : The energy reserves which are stored in a cell when there is excess of carbon available.

Strain : Another classification that refers to plants, usually annuals, that have similar growth characteristics but don't necessarily look alike.

Strandings : Aquatic mammals that get stranded on the beaches or shores.

Streptophytes : The clade consisting of the plants plus their closest relatives, the charophytes.

Stress : Any factor that potentially hinders an organisms development and life cycle.

Stria : A fine longitudinal line or ridge; hence striate (pl. striae).

Striate : Marked with fine longitudinal lines or ridges.

Strict : Very straight and upright.

Strigose : Covered with rough, stiff, sharp hairs that are more or less parallel to a particular surface.

Strobilus (cone) : Fertile stem with short inter nodes and sporophylls bearing sporangia, for example, as in many species of Lycopodium.

Stroma : The fluid filled area outside of the thylakoid involved in synthesising organic molecules from carbon dioxide and water.

Stroma Reactions : These are the set of reactions which occur in the stroma during the process of photosynthesis, without being directly powered by light.

Stromatolite : The stromatolite is a biological fossil representing colonies of bacteria, usually cyanobacteria, alternating with sediment layers.

Structural Genomics : The effort to determine the 3D structures of large numbers of proteins using both experimental techniques and computer.

Structural Region : It is the part of the gene, comprising of nucleotide triplets, that specifies which amino acids are to be incorporated into protein.

Style : In plants, style is a narrow elongated part of the pistil, located between the ovary and the stigma.

Sub : A prefix applied to many botanical terms meaning more or less, or somewhat; below.

Subelliptical : An egg that is elongated and tapered towards its rounded ends.

Subfamily : A grouping of similar genera within a family, a category above tribe (viz.).

Subsericeous : Covered with appressed hairs aligned in the same direction, but lacking a lustrous sheen; like silky but coarser.

Subshrub : Under shrub; a small and sometimes sparsely branched woody plant less than 1 m high.

Subspecies : A taxonomic category below species, differing in minor morphological characters such as size or shape of parts, and either partially or completely isolated, usually by means of geographical or ecological barriers.

Substitution : In genetics, a type of mutation due to replacement of one nucleotide in a DNA sequence by another nucleotide or replacement of one amino acid in a protein by another amino acid.

Substrate : A base on which an organism is grown. They can also be the substances on which compounds and enzymes act.

Substrate Level Phosphorylation : It is the formation of adenosine triphosphate (ATP) from adenosine diphosphate (ADP) by transferring a phosphate group from a substrate molecule.

Substrate Specificality : In botany, the term substrate specificality is used to refer to the ability of a given enzyme to distinguish one substrate from other similar substrates.

Subtend : To occur immediately below as in a bract subtending a flower.

Subtended : Lying under or opposite to.

Subtending : Term describing a leaf or bract whose axil gives rise to a bud (the axillary bud) which may develop into a branch or inflorescence; less commonly (as in Notelaea species) more than one bud is subtended in each axil.

Subulate : Awl-shaped; tapering from a thicker base to a sharp point, the sides usually concave.

Succulent : Fleshy, juicy and thickened.

Sucker : Growth that occurs from the root stock rather than from the grafted region. For example, non-disease resistant roses are often grafted to a disease resistant root stock. Without proper maintenance suckers will grow from the root stock.

Suction Disks : A type of attachment adaptation on certain types of vines.

Suffruticose : Low shrubby, with the lower part of the stem woody and the upper part herbaceous.

Sulcus : A furrow or groove (pl. sulci.); hence sulcate, longitudinally grooved.

Sulfur (S) : Often used as a control for powdery mildew and rust. It is also a nutrient for plants that acts with nitrogen to produce protoplasm in plant cells.

Sulfur Cycle : The cycle wherein sulfur, the element is taken up by living organisms, then released upon the death of the organism, and then converted to its final state of oxidation.

Summer Annual : Plant with seeds germinating in spring or early summer and completing flowering and fruiting in late summer or early fall.

Superficial : On the surface, away from the margin.

Superior : Inserted above another organ or part; a superior ovary is free from the receptacle, with the perianth and stamens inserted below it or on a perigynous hypanthium.

Superior Ovary : One that is located above the perianth and free of it.

Superposed : Placed vertically over some other part.

Superposed Bud : An accessory bud found above the main axillary bud.

Supporting Cell : Cell bearing the carpogonial branch in Rhodophyta.

Suppressor Gene : A gene that can suppress the action of another gene.

Supra : Axillary; located above an axil.

Suprafamilial : Above the rank of family; similarly supra generic etc.

Surcurrent : Extending upward from the point of insertion, as a leaf base that extends up along the stem.

Surficial : Growing near the ground, or spread over the surface of the ground.

Suspensor : The suspensor is the cell or filament supporting a gamete, most often observed in a zygospore.

Suture : A junction or seam of union; a line of opening or dehiscence.

Swale : A depression or shallow hollow in the ground, typically moist.

Syconium : A 'fig', the multiple fruit formed in figs (Ficus species) by the invagination of the floral axis where the minute flowers and fruits are actually inside the swollen inflorescence stem.

Symbiosis : Two dissimilar organisms, living together. Their association maybe commensal or mutualistic.

Symmetric : Divisible into two or more equal parts.

Symmetrical : Balanced; with mirror image parts.

Sympatric : Of two or more species having coincident or overlapping ranges of geographic distribution.

Sympatric Speciation : The speciation, i.e. the evolution of a biological species, which occurs within a limited geographical area is known as sympatric speciation.

Sympetalous : Having the petals more or less united.

Symplast : The inner side of the plasma membrane, wherein water and low molecular solutes diffuse freely is known as the symplast of the plant.

Sympodial : Mode of development where primary axis is continually replace by lateral axes which become temporarily dominant then are replaced by their own laterals.

Sympodium : An axis made up of the basal portions of several branches, the apex of each branch dying and growth continued by an axillary bud; hence sympodial.

Syn : A prefix meaning 'together'.

Synangium : A sorus-like structure formed by the fusion of two or more sporangia, a single body divided internally into several loculus each bearing spores.

Synapsis : The pairing of two homologous chromosomes which occurs during the process of meiosis is known as synapsis.

Syncarp : A multiple fruit consisting of several united fruits, originating from several originally free carpels, usually fleshy.

Syncarpous : A gynoecium consisting of a number of carpels in which at least the ovaries are united; the ovary is then said to be compound.

Syndrome : The group or recognizable pattern of symptoms or abnormalities that indicate a particular trait or disease.

Synergism : Association between two organisms that is mutually beneficial.

Syngamy : Syngamy is the process fusion of a sperm and an egg.

Syngeneic : Genetically identical members of the same species.

Synoecious : Having male and female flowers in the same flowerhead.

Synonym : Another name for the same taxon, either the result of independent description, or the transference or combination of different taxa.

Synopsis : The pairing of two homologous chromosomes which occurs during the process of meiosis is known as synopsis.

Synsepalous : Having the sepals more or less united.

Synteny : Genes occurring in the same order on chromosomes of different species.

Syntrophy : Interaction between two or more populations that supply each other's nutritional needs.

Syntype : One of two or more specimens used by the author without designating a holotype for one of two or more specimens designated as the type.

Systemic : Something that involves the entire body and is not localised in the body.

T

Taiga : Coniferous evergreen forests found in the south of the tundra and north temperate region, characterised by harsh winters.

Tail Slapping : The forceful slapping of tails on the surface of water by dolphins.

Tansfection : The introduction of foreign DNA into a host cell.

Taproot : Large, vertical root with many smaller lateral roots. Taproots usually go deep into the soil and provide the plant a strong anchor. Taproots are characteristic of many dicots.

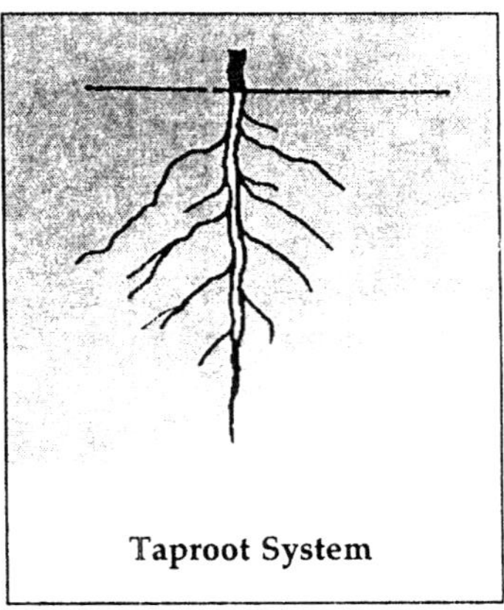

Taproot System

Targeted Mutagenesis : Deliberate change in the genetic structure directed at a specific site on the chromosome. Used in research to determine the targeted region's function.

Tarsus : The bone, which contributes in making the ankle joint, located between the tibia, fibula and metatarsus in mammals.

Tautonym : An in a dmissable name where the genus name is the same as the species name.

Taxis : The movement of a cell that is triggered by external stimulus, towards or away from the stimulus source, is known as taxis.

Taxon : A term used to describe a member of any taxonomic category, e.g. genus, species. pl. taxa.

Taxonomic Classification : The hierarchical system used for grouping and naming species of living organisms.

Taxonomic Synonyms : Synonyms (viz.) with different basionyms (viz.) or based on different types (viz.).

Taxonomy : A practice used to classify plants with evolutionary relationships, as basis of this classification.

Technology Transfer : The process of transferring scientific findings from research laboratories to the commercial sector.

Teichoic Acids : All wall, membrane or capsular polymers containing glycerophosphate or ribitol phosphate residues.

Telemorph : One of the stages of sexual reproduction, wherein cells are formed by meiosis and genetic recombination.

Telomerase : The enzyme that directs the replication of telomeres.

Telomere : The end of a chromosome. This specialised structure is involved in the replication and stability of linear DNA molecules.

Temperate : Pertaining to the cooler areas of the world with moderate climates.

Temperate Virus : A virus that does not cause destruction and lysis of the cells of its host, but instead, its genome may replicate in sync with that of the host.

Tendril : Tendril is a narrow stem-like structure which helps the twining plants in attaching themselves to an object in order to gain support from it.

Tepal : When the sepals and petals of a flower are indistinguishable, they are referred to as tepals. Tepals are common in many groups of monocots.

Teratogenic : Substances such as chemicals or radiation that cause abnormal development of a embryo.

Terete : Cylindrical, circular in transverse section.

Terminal : At the end of the branch or stem.

Terminal Bud : The "tip" or "end" of a stem, usually where new plant growth is concentrated.

Terminal Electron Acceptor : The last acceptor of the electron, as it exits the electron transport chain.

Terminal Petiolule : The stalk of the terminal leaflet of a pinnately 3-foliolate leaf or an imparipinnate leaf; the stalk is usually jointed at the point where the rachis extension beyond the last leaflet meets the true petiolule of the leaflet.

Ternate : In threes, e.g. of a single leaf, having the leaflets arranged in groups of three.

Terrestrial : Of the land as opposed to living in water.

Tesselate : Of a checker-work pattern or divided into squares or squarish pieces.

Tessellate : With colours or shapes arranged in squares to give a chequered appearance, e.g. of bark.

Test Cross : The test cross is a process wherein a suspected heterozygote is tested by crossing it with a known homozygous recessive.

Tetra : Prefix meaning four.

Tetrad : A group of four; as in four pollen grains remaining together at maturity in Epacridaceae.

Tetrahedral : Of a solid with four sides.

Tetrasporangia : Cell in which diploid (2N) nucleus undergoes meiosis and forms four haploid (N) spores or tetraspores in the Rhodophyta.

Tetrasporophyte : Diploid phase which produces tetraspores in the Rhodophyta.

Thalloid : Plants which have no roots, stems, or leaves are called thalloid, such as liverworts and hornworts.

Thallus : Thallus is a plant which doesn't feature true stems, roots, leaves or vascular system.

Thermocline : That point in a lake, where there is a drastic drop in temperature with increase in depth.

Thermophile : An organism that grows best at temperatures around 45 and 80 degrees Celsius.

Thorax : The part of the body in mammals situated between the neck and the abdomen, just above the diaphragm. In case of insects the part situated between the head and the abdomen, excluding the legs and wings.

Thorn : Thorns, also referred to as spines, are the leaves of plants which are modified into cylindrical, hard structures featuring sharp ends.

Threatened Species : A species which has a possibility of becoming endangered in the near future.

Thylakoid : The thylakoid is a membrane-bound compartment within the chloroplasts and cyanobacteria which is a site for the light-dependent reactions of photosynthesis.

Thymine (T) : A nitrogenous base, one member of the base pair AT (adenine-thymine).

Thyrse : A compound inflorescence ending in a vegetative (non-floral) bud and with mixed types of branching, the main axis bearing several or many lateral cymes.

Thyrsiform : An inflorescence shaped like that of a lilac or bunch of grapes.

Thyrsoid : A compound inflorescence which ends in a flower and in which the main axis is raceme-like and the lateral ones cymose, i.e. similar to a thyrse except for the terminal flower.

Ti Plasmid : A conjugative tumour inducing plasmid that can transfer genes into plants. Seen in the bacterium *Agrobacterium tunefaciens*.

Tiller : In grasses the young vegetative shoots.

Tinsel Flagellum : A flagellum which is covered with several minuscule hair like projections is referred to as tinsel flagellum.

Tip Layering : Tip layering is a plant propagation method wherein only the stem tip is buried in order to facilitate the growth of a new plant.

Tissue : A tissue is an ensemble of cells featuring similar structure and performing a specific function.

Tissue Culture : Tissue culture is a process wherein various cells are separated from each other and grown outside the body, on a culture medium.

Tomentose : Woolly, with long, soft, matted hairs.

Tomentum : A dense woolly or matted covering of +/- appressed hairs; hence tomentose.

Toothed : Furnished with short projections, especially when these are sharp.

Topotype : A specimen collected from near the locality of the type (viz.).

Torsion : The asymmetrical positioning of the body achieved, due to twisting and repositioning, during development.

Toxicogenomics : The study of how genomes respond to environmental stressors or toxicants. Combines genome-wide mRNA expression profiling with protein expression patterns using bioinformatics to understand the role of gene-environment interactions in disease and dysfunction.

Toxin : A foreign substance present in the body, which is mostly generated by microorganisms, that is capable of inflicting damage on the host cell.

Trabecula : A transverse partition dividing or partly dividing a cavity; hence trabeculate.

Tracheid : The elongated cells in the xylem which facilitate the transportation of water and mineral salts within the plants are known as tracheids.

Tracheophyte : Any member of the clade of plants possessing vascular tissue; a vascular plant.

Transcription : A process facilitated by the enzymes to transcribe the information of a DNA strand into a complementary RNA(tRNA) strand is known as transcription.

Transcription Factor : A protein that binds to regulatory regions and helps control gene expression.

Transcriptome : The full complement of activated genes, mRNAs, or transcripts in a particular tissue at a particular time.

Transduction : The process where host genetic information is transferred through an agent like a virus or a bacteriophage.

Transfer RNA (tRNA) : A class of RNA having structures with triplet nucleotide sequences that are complementary to the triplet nucleotide coding sequences of mRNA. The role of tRNAs in protein synthesis is to bond with amino acids and transfer them to the ribosomes, where proteins are assembled according to the genetic code carried by mRNA.

Transformation : A process by which the genetic material carried by an individual cell is altered by incorporation of exogenous DNA into its genome.

Transgenic : An experimentally produced organism in which DNA has been artificially introduced and incorporated into the organism's germ line.

Transgenic Plant : A plant which contains DNA inserted by some form of genetic engineering is known as transgenic plant.

Transition Zone : Intercalary meristem in region between stipe and blade where active cell division occurs in the brown algal order Laminariales.

Translation : The term translation is used to refer to a process wherein the sequence of amino acids is facilitated during protein synthesis by the information in an mRNA strand.

Translocation : The process of transportation of dissolved material within a plant is referred to as translocation.

Translucent : Letting light through, almost transparent.

Transpiration : In botanical studies, the process of emission of water vapour from the plant leaves is known as transpiration.

Transposable Element : A class of DNA sequences that can move from one chromosomal site to another.

Transposition : A form of chromosomal mutation wherein a chromosomal segment is transferred to a new position on the same or some other chromosome.

Transposon : Transposable element which, in addition to transposable genes, carries other genes.

Transposon Mutagenesis : A mutant phenotype is formed by inactivation of the host gene, which occurs due to the insertion of a transposon.

Transposons : The sequences of DNA which can move to different positions within the genome of a single cell through the process of transposition.

Transverse : At a right angle to the longitudinal axis of a structure.

Trapezoid : Trapezium-shaped, a four-sided figure with the opposite sides parallel.

Tree : Any tall plant, including many conifers and flowering plants, as well as extinct lycophytes and sphenophytes.

Tri : Prefix meaning three.

Triad : A cluster of three, as spikelets of *Hordeum* or *Hilaria.*

Triandrous : Having three stamens.

Triangular : A 2-dimensional shape, 3-angled and 3-sided.

Tribe : A category in the classification of organisms between a genus, which contains one or more genera.

Tricarboxylic Acid Cycle : A series of metabolic reactions, by which pyruvate is oxidised to carbon dioxide.

Trichoblast : Simple or branched filament at apex of the red family Rhodomelaceae.

Trichogyne : Receptive elongation of the female reproductive structure in the Rhodophyta where male gametes become attached.

Trichome : Trichomes are the various extensions developing from the epidermis of the plant which are meant to provide protection to the plant.

Trichothallic Growth : Growth from cell division at the base of a filament or group of filaments as in the brown algal order Desmarestiales.

Tridentate : Having three teeth, prongs or points.

Trifid : Divided into three +/- equal parts, usually to about halfway.

Trifoliolate : Of a compound leaf with three radial leaflets.

Trifurcate : Split into three forks or branches.

Trigonous : With three angles, of a solid body, triangular in cross-section with rounded corners.

Tripartite : Divided into three +/- equal parts, to the base or almost so.

Tripinnate Leaf (3-pinnate) : A compound leaf with lamina pinnately divided three times, i.e. the pinnules are again pinnately divided.

Tri-pinnately Compound : Thrice pinnately compound.

Triquetrous : Triangular in cross-section and sharply angled; with three distinct longitudinal ridges.

Trisomy : Possessing three copies of a particular chromosome instead of the normal two copies.

Tristylous : Heterostylous species with styles of 3 different lengths (short, mid, long).

Tropic : Pertaining to the warmer or equatorial regions of the world; hence tropical.

Tropism : A biological process, which indicates the growth of a plant, in response to the environmental stimulus is known as tropism.

Truncate : With an abruptly transverse edge as if cut off, e.g. of a lamina apex, or base.

Trunk : The erect, unbranched portion of a treelike plant.

Tube **:** Any hollow elongate body of an organ; hence tubular.

Tubenoses **:** Vernacular name for members belonging to Procellariiformes species.

Tuber **:** The various types of modified plant structures which are enlarged to store nutrients are known as tubers.

Tubercle **:** A small wartlike outgrowth, e.g. forming the base of a hair.

Tubercle, Tubercule **:** A small, wartlike swelling.

Tuberculate (Warty) **:** An organism or part of an organism which is covered in fleshy and raised protuberances, also called tubercules.

Tuberous Root **:** A thickened true root used to store nutrients. An example is dahlia.

Tubulin **:** Tubulin is a protein which leads to formation of microtubules on polymerisation.

Tumescent **:** Somewhat tumid, swelling.

Tunicate **:** Having several concentric layers, such as in onions.

Turbinate **:** Shaped like a top or inverted cone.

Turgid **:** In botany, the word turgid is used to refer to a plant with swollen tissues which are filled with moisture.

Turgor Pressure **:** The outward pressure exerted by the water in the plant cells, which adds to the rigidity of these cells, is known as the turgor pressure.

Twig **:** A young shoot of the past season.

Twiner **:** A climbing plant supporting itself by winding spirally around an object.

Twining **:** Climbing by coiling around some support.

Two-ranked (2-ranked) **:** Arranged in two rows on opposite sides of a stem and in the same plane.

Tylosis **:** The tylosis is the process wherein an outgrowth from a parenchyma cell, through the pit cavity into a vessel, leads to the blockage of the vessel.

Tympanic Membrane : It is the membrane which picks up vibrations through a medium and transports them to the inner part of the ear. It is also called the 'eardrum'.

Type : The designated representative of a taxon constituting a fixed point for the application of its name, for determining priority of usage.

Type Genus : The genus to which the name of a genus is permanently attached.

Type Species : The species to which the name of a genus is permanently attached.

Type Specimen : An organism which is used to represent a particular taxon. It becomes the standard for the original name and to describe the species.

Type Subspecies, Type Variety : The variety or subspecies of the species that contains the nomenclatural type of the species.

Type, Type Specimen : The nomenclatural type or specimen to which the name of a taxon is always attached.

U

Umbel **:** A flower cluster where the pedicels arise from a common point and the pedicels are of nearly equal length, forming an umbrellalike inflorescence.

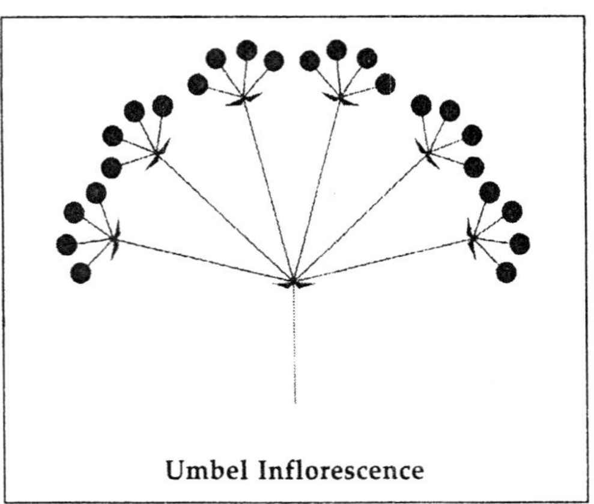

Umbel Inflorescence

Umbellaster **:** A group of flowers (with the terminal bud ending in a flower) more or less arising from the one point, which may be derived from the condensation of a cyme, thyrsoid or panicle.

Umbellet **:** A secondary umbel in a compound umbel.

Umbellulate **:** In the form of or having the appearance of an umbel.

Umbo **:** A rounded protuberance or elevation in the centre; hence umbonate, bearing a protuberance in the centre.

Umbrel : A type of inflorescence in which all of the individual flowers arise from the same point. For example, the flowers of Yarrow.

Unarmed : Without spines, prickles and the like.

Uncinate : Hooked near the apex or having the form of a hook.

Underplanting : Planting plants beneath other plants. For example planting ground cover plants beneath a tree.

Undershrub : A small shrub, often partially herbaceous.

Undulate : Corrugated or with the margin waved in a plane perpendicular to the surface.

Unguiculate : Contracted at the base into a claw, as a petal.

Uni : Prefix meaning 'one' or 'solitary'.

Uniflorescence : A unit inflorescence forming part of a conflorescence.

Unifoliolate Leaf (1-foliolate) : A compound leaf reduced to a single leaflet, usually recognised by the articulated or jointed 'petiole', which is in fact a petiole plus a petiolule.

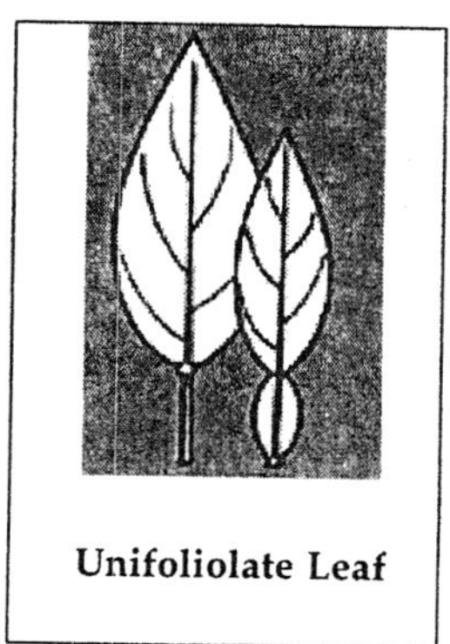

Unifoliolate Leaf

Unigeneric : Of a family, having only one genus.

Unilocular : Produced in a single cavity or cell as in the unilocular sporangium of the brown algal order Ectocarpales.

Univarental Inheritance : Genetic inheritance obtained from just one parent, and is generally the case for mitochondrial and plastid genes.

Uniseriate : Arranged in one row or series.

Unisexual : Flowers that have either the pistil or the stamen are referred to as unisexual flowers.

Uracil : A nitrogenous base normally found in RNA but not DNA; uracil is capable of forming a base pair with adenine.

Urceolate : Hollow and contracted at the mouth like an urn or pitcher.

Uronic Acid : A class of acidic compounds that contain both carboxylic and aldehydic groups and are oxidation products of sugars. They occur mainly in polysaccharides.

Utricle : Swollen terminal portion of filament or siphon as in siphonous green alga *Codium*.

Uva : A grapelike berry formed from a superior ovary.

V

Vacuolar Membrane : The membrane sheathing the cell vacuole.

Vacuole : Fluid pocket separated from the cell cytoplasm by a membrane, which mostly occupies about 99% of the cell's volume, and stores dissolved matter.

Vadose Zone : Unsaturated zone of soil which is above the ground-water, extending from the bottom of the capillary fringe to the soil surface.

Vaginate : Provided with or surrounded by a sheath.

Vagrant : An individual organism found outside the region that is known for that particular species.

Valvate : (1) opening by valves, e.g. loculicidal andsepticidal capsules, or of anther dehiscence; (2) of floral parts, with the edges touching but not overlapping.

Valve : One of the pieces into which a dehiscent pod, or any similar structure, splits; one of the parts into which a capsule splits.

Variegation : The presence of another colour other than the main colour on the leaf, stem, or petal of a plant. For example, green with white stripes on leaves, or green leaves with white edging on ivy.

Vascular : Containing both xylem, the principal water and mineral-conducting tissue, and phloem, food conducting tissue.

Vascular Bundle : Column of tissue comprising mostly of phloem and xylem, which are usually enveloped by a bundle sheath.

Vascular Cambium : Meristem present in the form of narrow cylindrical sheath, that produces secondary xylem and phloem in the roots and stems.

Vascular Plant : Plants possessing the vascular tissues i.e. xylem and phloem are termed as vascular plants.

Vascular Tissue : Tissue formed into tubes within a plant that transports water and nutrients throughout the plant.

Vascular Tissue System : The system formed from xylem and phloem to transport water and nutrients throughout the plant.

Vector : An agent that can carry pathogens from one host to another. It can also denote a plasmid or virus used in genetic engineering to insert genes into a cell.

Vegetation : The total aggregation of plant communities within an area.

Vegetative Cell : A growing or actively feeding form of a cell, as against a spore.

Vegetative Growth : Growth of a plant by division of cells, without sexual reproduction.

Velamen : A water-retaining outer layer of aerial roots of some epiphytes, especially orchids.

Velamen Root : Aerial root capable of preventing water loss due to its multilayered epidermis.

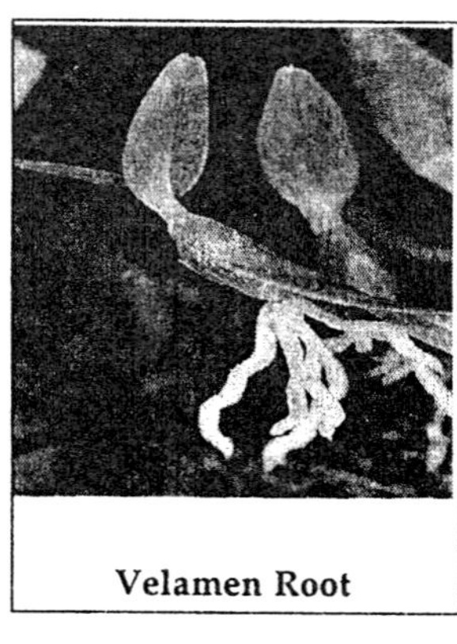

Velamen Root

Velvety : Very densely covered with fine short soft erect hairs.

Venation : The arrangement of the veins, especially in leaves or leaflike structures.

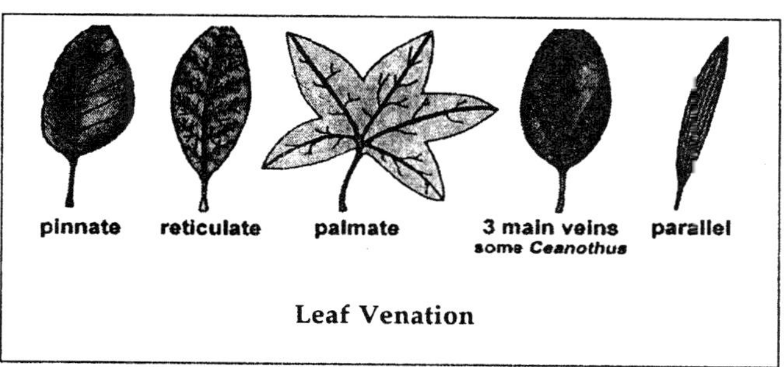

Leaf Venation

Venter : Egg's site in the large basal region of the archegonium.

Ventral : On the inner or axis side of an organ or the upper surface of a leaf.

Ventral Scale : Scales found on the underside of the snake's body.

Ventricose : Inflated or swollen unequally on one side.

Vermicular : Worm-shaped or wormlike, or of worm-eaten appearance.

Vernalisation : Cold treatment required to initiate flowering in biennials.

Vernation : The arrangement of unexpended leaves in a vegetative bud.

Vernicose : Appearing as though varnished.

Verrucose : With a warty, nodular surface (dimin. verriculose).

Versatile : Referring to an anther which attaches at or near its middle and is able to turn freely on its support.

Versicolour : Having various colours.

Verticellate : In a circle or whorled about the axis.

Vesicle : A bladderlike sac or cavity filled with gas or liquid. adj. vesicular, e.g. of hairs that are inflated and bladderlike; vesicular hairs often collapse and form a silvery layer on the surface of the organ on which they are formed.

W

Warty : Covered with wartlike protuberances.

Water Content : The amount of water contained in a material, which is expressed as the mass of water per unit mass of oven: dry material.

Water Potential : Amount of water that can be absorbed or released by a substance with respective to another substance is termed as water potential.

Water Retention Curve : A graph showing soil water content as a function of increasingly negative soil water potential.

Water Sprout : Also known as a sucker, a vertical shoot growing from the main trunk or branches.

Water Vascular System : A system of fluid filled tubes and ducts, that connect with the tube feet of most marine invertebrates. They help in functions of respiration, feeding, etc.

Water-splitting : Phenomenon occurring in photosystem II of the process of photosynthesis, wherein water molecules split to release oxygen.

Weaning : The period where the mother ceases to feed the young ones. This only refers to mammals.

Webbing : According to telome theory of megaphyll origin, the lamina originated from parenchymatic cell production between the telomes.

Weed : Any unwanted plant in a garden or wildlands. Typically pests that regenerate themselves. For example, dandelions, scarlet pimpernel, etc. Weeds can provide beneficial nutrients to the soil.

other and with grasses and other herbs forming a more or less continuous ground cover between them.

Woody : Refers to plants with hardened wood trunks.

Woolly : Densely covered with matted long hairs.

Woronin Body : A spherical structure found in fungi belonging to the phylum *Ascomycota*, which are associated with the simple pore in the septa separating the hyphal compartments.

Wrinkled : Creased, folded up irregularly in every direction.

X

X : A symbol which when placed before a specific epithet indicates a hybrid of two species.

X Chromosome : One of the two sex chromosomes, X and Y.

Xenobiotic : A compound that is foreign to the biological systems.

Xenophobic Alliance : A union of individual chimpanzees in a group, which challenges intruders who threaten their territory and boundaries.

Xeric : A habitat which has an extremely dry environment.

Xerophile : An organism that is capable of growing at low water potentials, that is, in very dry habitats.

Xerophilous : Growing in dry places.

Xerophyte : A plant which has adapted itself to a dry environment and is able to conserve water.

Xerophytic : Adapted to dry or arid conditions, places where fresh water is scarce or where water absorption is difficult due to an excess of dissolved salts.

Xylem : The portion of conducting vascular tissue that conducts water and dissolved minerals. It contains several types of cells such as tracheids, vessel elements, parenchyma, sclereids, fibres, etc.

Xylocarp : A hard, woody fruit such as the coconut.

Xylophagous : Organisms that feed entirely or primarily on wood.

Y

Y Chromosome **:** One of the two sex chromosomes, X and Y.

Yearling **:** This term is used to describe both a male and female horse between the age of one and two years.

Yeast Artificial Chromosome (YAC) **:** Constructed from yeast DNA, it is a vector used to clone large DNA fragments.

Yeasts **:** Unicellular ascomycetes which lack mycelium. In yeasts individual cells itself perform the functions of a large mycelium.

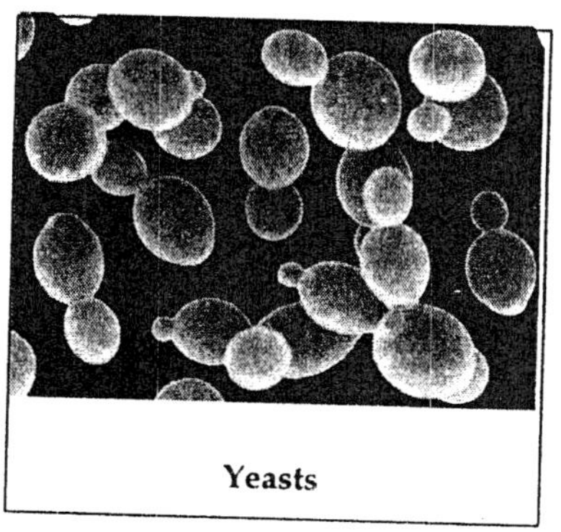

Yeasts

Z

Zinc-finger Protein : A secondary feature of some proteins containing a zinc atom; a DNA-binding protein.

Zone of Elongation : Root tip region which lies toward the root apical mersitem, where pronounced elongation of cells takes place.

Zooplankton : A collection of various species of plankton.

Zoospore : Motile spore capable of swimming. Occurs in fungi and algae.

Zooxanthella : Symbiotic golden-pigmented algal cell in tissue of animal host.

Zosterophyllophytes : Bunch of early vascular plants possessing xylem and exarch prostele. They also feature lateral sporangia which open transversely on the top edge.

Zygomorphic : Of a flower with the parts such as sepals and petals differing in shape, size, position and/or number so that the flower can be bisected in one plane only; bilaterally symmetrical.

Zygosporangium : Large multinucleate sporangium produced by the fusion of two compatible hyphae in Zygomycete fungi.

Zygote : The fertilised egg before it undergoes differentiation; the result of the union between male and female gametes.

Zygotic Meiosis : Meiosis that occurs during zygote maturation or germination.

Zymogenous Flora : Refers to microorganisms that respond rapidly by enzyme production and growth when simple organic substrates become available.